This is Physics!

by Andrew D. Bunch

Copyright © 2021 Andrew D. Bunch
All rights reserved.
ISBN: 9798500464552
Independently Published through Amazon Kindle Direct Publishing

Thank you to Hamid Aidinejad for your friendship and mentoring, to Ken Shacter and William Mendoza for your very useful comments and edits, to all of my students and colleagues in my academic career for making it all worthwhile, and to my friends and family for your support over the years.

Most of all…
Thanks, Mom!
You're the best, Mom!
What would I do without you, Mom!

Table of Contents

Ch. 1 – Describing Motion p.5

Ch. 2 – Mass, Momentum, and Forces p.22

Ch. 3 – Energy p.45

Ch. 4 – Thermodynamics p.61

Ch. 5 – Fluid Principles p.80

Ch. 6 – Waves p.92

Ch. 7 – Electricity and Magnetism p.102

Ch. 8 – Circuits p.121

Ch. 9 – Optics p.143

Ch. 10 – Relativity p.162

Ch. 11 – Quantum mechanics p.176

Ch. 12 – Nuclear and Particle Physics p.192

Ch. 1 – Describing Motion

Getting Started

Physics is the science that describes how objects and systems interact on a fundamental level and how they evolve over time. As such, Physics covers a lot of things! What we will try to do over the course of this book is give you a rough overview of the main topics that form the backbone of the science – the stuff that every physicist learns when he or she gets started. To make it manageable, we'll keep the math at the high school level, with more advanced math in footnotes.

Physics is an observational science – the information presented here are ideas and models that accurately reflect and predict what we observe around us. For the sake of brevity, much of the history of its development will be omitted. One thing to keep in mind as we go forward is that although we use many physics terms in our everyday language, those terms mean very specific things in the science that may be somewhat different from their conventional uses. As each term is introduced, we'll go over the formal definition and the symbol that is used to represent it in equations. Unfortunately, there aren't enough letters in the alphabet, Greek or otherwise, for every term to have a unique identifier – context is key. Lastly, the units used in Physics are metric.

The typical starting point in physics is describing motion. To make things as simple as possible, we'll just treat an object as existing at a single point and ignore any rotation for now. The position or location of a particle can be represented in a Cartesian Coordinate system. A two-dimensional Cartesian Coordinate system with a highlighted point is given below.

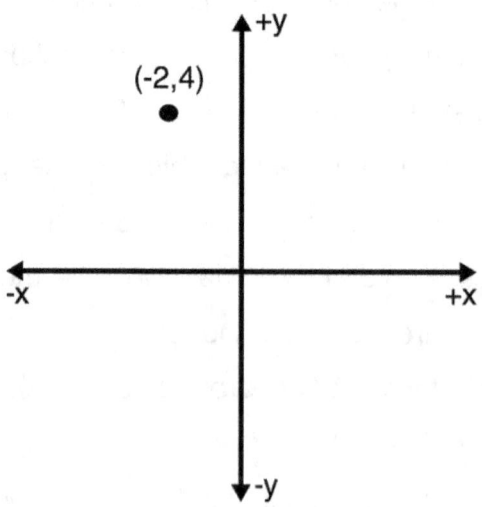

In this graph, the horizontal line is called the x-axis, and the vertical line is the
y-axis. The point drawn on the graph represents our object located at point (-2,4): -2 in the x-direction and +4 in the y-direction. The x and y axes can represent left/right and

up/down, north/south and west/east, or whatever seems appropriate for the situation.

The next step in talking about motion is describing how an object's position changes. Below is another graph that follows our object moving from our original point to another point.

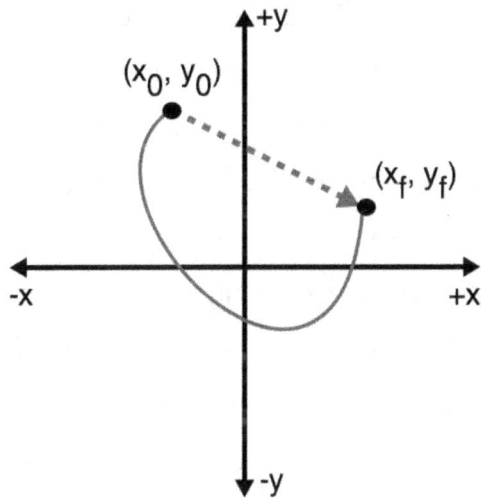

The original point is labeled (x_0, y_0), and the final point is labelled (x_f, y_f). In physics, we use the subscript "0" to represent the initial value of something, and the subscript "f" to represents its final value.

We can describe how the position changes in this graph using two different frameworks. The first framework is using a class of variables called **scalars**. Scalars are just numbers – they have some value, possibly zero or negative, and may or may not have units attached. The second framework is using a class of variables called **vectors**.

Vectors are variables that have multiple pieces of information in them. Vectors have both magnitude *and* direction. The downside for having this extra information is that vectors require more care when manipulating them mathematically. In many texts, vectors are either written in bold text or have an arrow over the variable's symbol.

The scalar term for describing how an object's position changes is **distance**, d. It is equal to the length of the path taken. In the previous graph, the solid curved line represents the path of the particle. This is what your odometer measures in a car. It doesn't matter which roads you take – the miles keep adding up as you go along.

The vector term for describing how an object's position changes is **displacement**, Δx and Δy. It is equal to the object's final position minus its initial position in each direction. This is akin to drawing a straight line from one point to another, as is done in the dashed line with the arrow in the previous graph. We use Δx to represent the change in the x-direction, and Δy for the change in the y-direction. The capital Greek letter delta, Δ, is used extensively in physics. It means "change in," or "final value minus initial value".

$$Change\ in\ x = Final\ x - Initial\ x$$
$$\Delta x = x_f - x_0$$

For both distance and displacement, the metric units used are meters, m.

To give another mathematical example, say that you travelled 40 km (kilometers) west and then 30 km south. The distance travelled would be 40 km plus 30 km, equaling 70 km total. If we were to look at the displacement, it would be 40 km west (Δx) and 30 km south (Δy). We could use the Pythagorean Theorem ($a^2 + b^2 = c^2$ for the sides of a right triangle) to find the length of the straight line going from the starting position to the ending position: $(40 \text{ km})^2 + (30 \text{ km})^2 = 2500 \text{ km}^2$. Taking the square route of the result, we find that the net displacement is 50 km southwest (at 37° south of west, to be precise, as seen in the following figure).

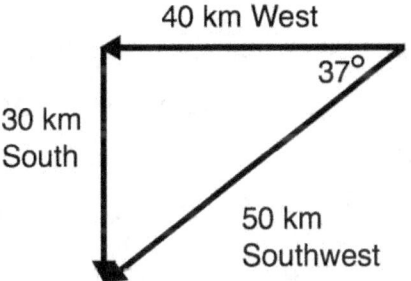

The next step is to look at how the position changes with time. This a vector quantity called the **velocity**, v, with base metric units of meters per second, m/s. Remember, vectors have multiple components, both magnitude and direction. The formal definition of the velocity is the change in position (displacement) divided by the change in time (elapsed time).

$$velocity = \frac{change\ in\ position}{change\ in\ time} = \frac{Final\ x - Initial\ x}{Final\ t - Initial\ t}$$

$$v = \frac{\Delta x}{\Delta t} = \frac{x_f - x_0}{t_f - t_0}$$

Frequently, the initial time can safely be set to zero, as is done with a stopwatch, and the initial position is often set to zero as well.

The magnitude of the velocity is a scalar called the **speed**, and is equal to the distance travelled divided by the elapsed time. The speedometer in a car only tells you the speed you are going – you need to have a compass if you want to know the direction, and thereby define the velocity. If the direction of motion never changes, velocity and speed tend to be used interchangeably.

Lastly, we can see how the velocity changes with the time, which is a vector called the **acceleration**, a, with units of meters per second squared, m/s² (meters per second per second). Formally, it is equal to the object's change in velocity divided by its change in time

$$acceleration = \frac{change\ in\ velocity}{change\ in\ time} = \frac{Final\ v - Initial\ v}{Final\ t - Initial\ t}$$

$$a = \frac{\Delta v}{\Delta t} = \frac{v_f - v_0}{t_f - t_0}$$

Unlike velocity, there isn't a special term used for the scalar form of the acceleration.

Now that we have given some formal definitions for terms used to describe motion, let's look at some important special cases and applications.

Motion at a Constant Velocity

The simplest example of motion is where an object is moving at a constant velocity, like a person walking at a steady pace or a car on cruise control on the highway. Since the velocity is constant, this means that the acceleration (the change in velocity over the change in time) is zero by default. Furthermore, if the velocity is constant, the position changes at a steady rate.

A normal walking speed is about 1.4 m/s. If a person walks at that constant speed for 10 seconds, that person has travelled a distance of 14 meters, which is fairly intuitive in this scenario. Here is how you go about solving that using our formal definitions from before. We are given the speed (or velocity, since we are only moving in one direction) to be 1.4 m/s (v = 1.4 m/s), and we are given the elapsed time of 10 seconds (Δt = 10 seconds). We just need to rearrange our equation for the velocity to solve for the change in position:

$$v = \frac{\Delta x}{\Delta t} \quad \rightarrow \quad \Delta x = v \cdot \Delta t = 1.4 \frac{m}{s} \cdot 10 \, s = 14 \, m$$

Frequently, physicists use graphs to help analyze how one variable influences another. Below is a plot of the person's position versus time. You can see how the position steadily increases due to that forward or positive velocity.

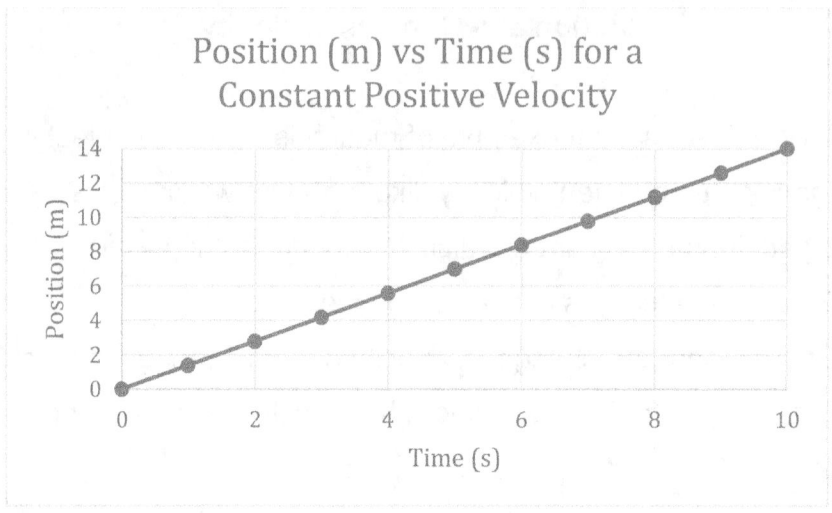

Motion with a Constant Linear Acceleration

The next step is to introduce some acceleration along a straight line. Let us say that you are in a car that speeds up, and to make the scenario as simple as possible, we will assume that the car speeds up or accelerates at a constant rate. The car is initially moving at 10 m/s, and it accelerates at a constant rate of 2.5 m/s² for eight seconds. How fast is the car moving after this time, and what is going on with the car's position?

First, let us apply our definition for the acceleration, which is the change in velocity divided by the change in time. We have the acceleration (a = 2.5 m/s²) and the elapsed time (Δt = 8 s). We can rearrange this equation to solve for the object's change in velocity.

$$a = \frac{\Delta v}{\Delta t} \rightarrow \Delta v = a \cdot \Delta t = 2.5 \frac{m}{s^2} \cdot 8\,s = 20 \frac{m}{s}$$

Since the car is initially moving at 10 m/s, and it experiences a change in velocity of 20 m/s, this means that the car is moving 30 m/s at the end of that acceleration. Below is a graph of the car's velocity versus time.

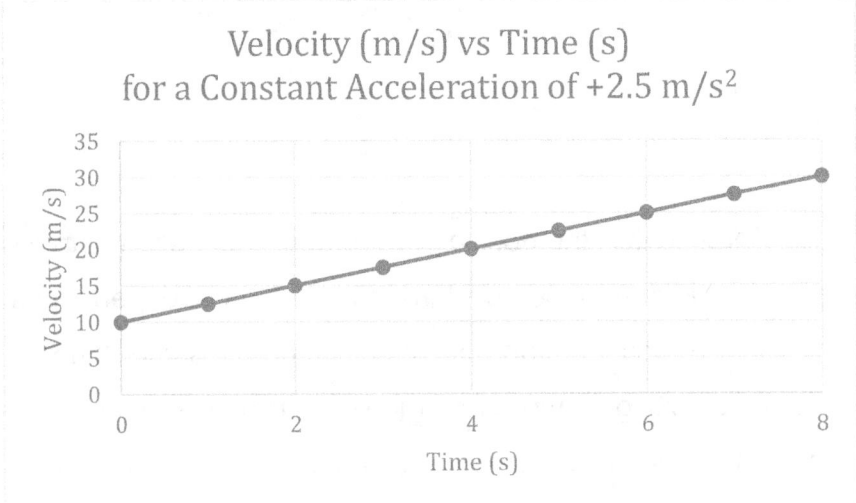

Even though this looks like the previous graph, the two graphs represent two different variables in two very different scenarios. The position graph for this case, instead, looks like this.

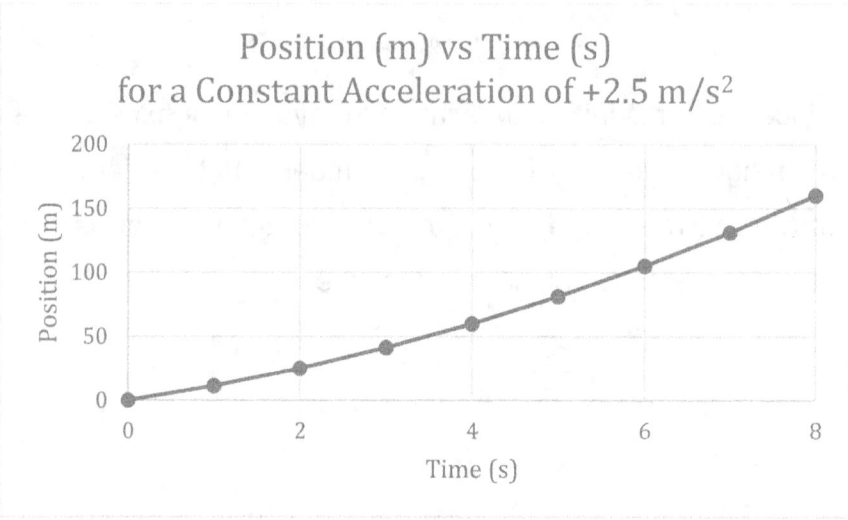

Notice that the position graph now has a slight curve to it. This is due to the acceleration. As the car experiences this positive acceleration, the velocity increases over time. As the velocity gets more and more positive, this means the position increases at a faster and faster rate. In other words, the at the moment you start accelerating, you are not moving very quickly, and are not increasing your position very much that first second. As time goes on, you're moving faster and faster, and for every second, you are covering a greater and greater distance.

If you want to solve for that final distance, first you can solve for your average speed, then multiply that by the elapsed time, just like what was done in the first problem given*. In this case, the average speed is 20 m/s, which we find as follows. The sum of the initial and final speeds is 30 m/s + 10 m/s = 40 m/s, and that sum divided by two gives the average of 20 m/s. Multiplying that average speed by the elapsed time of 8 seconds gives us the change in position of 160 m.

$$v = \frac{\Delta x}{\Delta t} \quad \rightarrow \quad \Delta x = v \cdot \Delta t = 20\frac{m}{s} \cdot 8\,s = 160\,m$$

Now let's look at a case where the acceleration is negative. We left off with the car moving at 30 m/s. The driver hits the breaks, and the car comes to a complete stop after three seconds. What was the acceleration during this time, and how far did it travel?

Again, we can use our definition of the acceleration, being the change in the object's velocity divided by its change in time. The car's final velocity is zero, since it came to a stop, and its initial velocity was 30 m/s. The elapsed time or change in time was given to be three seconds. Therefore, the car's acceleration is -10 m/s².

$$a = \frac{\Delta v}{\Delta t} = \frac{0 - 30\frac{m}{s}}{3\,s} = -10\frac{m}{s^2}$$

The graph for the car's velocity during this time is given below.

---

* Another equation that you can use to find the final position for an object undergoing a constant linear acceleration is as follows: $x_f = x_0 + v_0 t + \frac{1}{2}at^2$

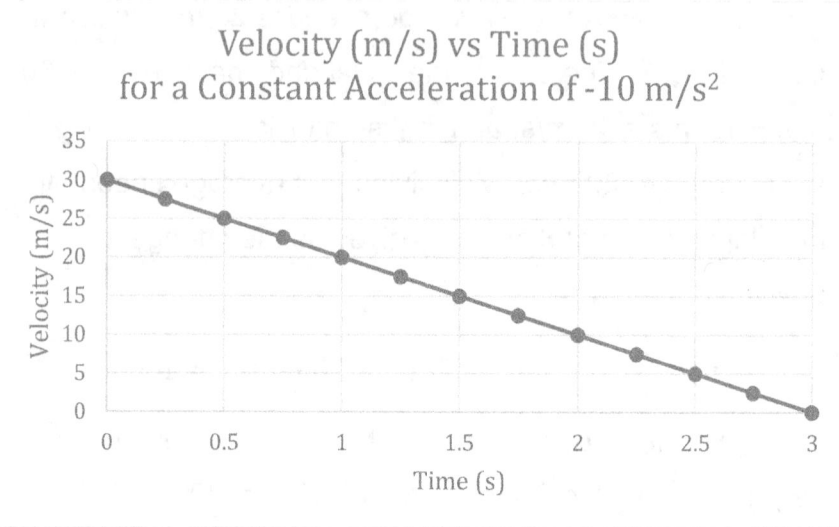

Lastly, let us look at the car's position versus time.

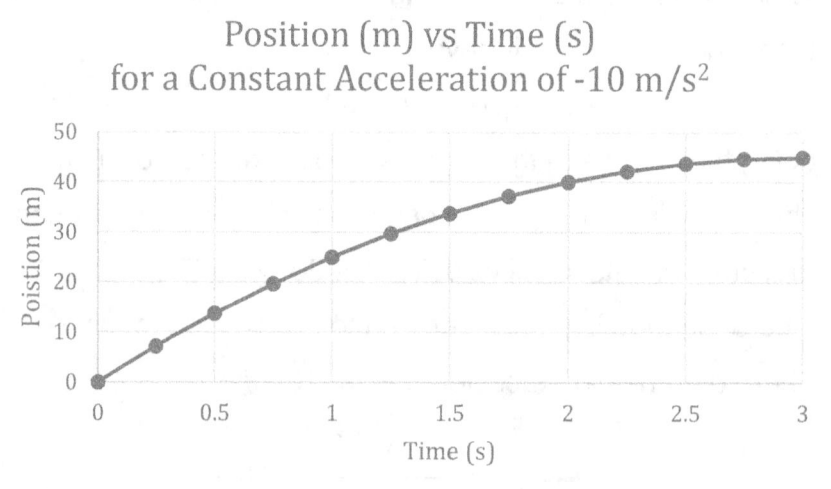

Here, due to the negative acceleration, the velocity is decreasing. Since the velocity is still positive (the car is still moving forward!), the position is still increasing. As the velocity becomes a smaller positive value, the position increases at a slower and slower rate. We can find the final position as we did before. First, we find the average speed during that time: 0 m/s + 30 m/s = 30 m/s, and divided by 2 gives us 15 m/s. Then, we multiply that by the elapsed time of 3 s. This gives our final position to be 45 m, which is the distance it took for the car to come to a complete stop.

Rotational Motion

The descriptions and formulations described previously can be applied for rotational motion as well. Instead of tracking how one's position, x, varies with time, we look at one's angle, θ, which is the lower-case Greek letter theta. An object's change in angle divided by its change in time is equal to its **angular velocity**, ω, which is the lower-case Greek letter omega.

$$angular\ velocity = \frac{change\ in\ angle}{change\ in\ time} = \frac{Final\ \theta - Initial\ \theta}{Final\ t - Initial\ t}$$

$$\omega = \frac{\Delta \theta}{\Delta t} = \frac{\theta_f - \theta_0}{t_f - t_0}$$

In Physics, when doing any sort of math involving rotation, we have to measure angles in radians, not degrees, so as to avoid unnecessary conversions for forces and torques[*]. One complete circle or revolution is 360 degrees or 2π radians, so

---

[*] This is because radians are actually unitless! Degrees are a set of units tacked onto angles.

1° = 0.0175 radians. Accordingly, when we calculate angular velocity, the fundamental unit is in radians per second (rad/s), not revolutions per minute (rpm), where 1 rad/s = 9.55 rpm.

Lastly, just as linear acceleration is an object's change in linear velocity divided by the change in time, an object's **angular acceleration** is its change in angular velocity divided by the change in time. The units for angular acceleration are radians per second squared, rad/s², and its symbol is α, the lower-case Greek letter alpha.

$$angular\ acceleration = \frac{change\ in\ angular\ velocity}{change\ in\ time}$$

$$= \frac{Final\ \omega - Initial\ \omega}{Final\ t - Initial\ t}$$

$$\alpha = \frac{\Delta \omega}{\Delta t} = \frac{\omega_f - \omega_0}{t_f - t_0}$$

Uniform Circular Motion

The last special case we'll consider in this chapter is that of uniform circular motion – an object moving in a perfect circle of radius, r, with a constant angular speed. In order for this to happen, the object must experience a constant acceleration towards the center of the circle (centripetal). Additionally, the object's linear velocity has a constant magnitude but its direction is always tangent to the circle. These quantities and their directions are depicted in the figure below.

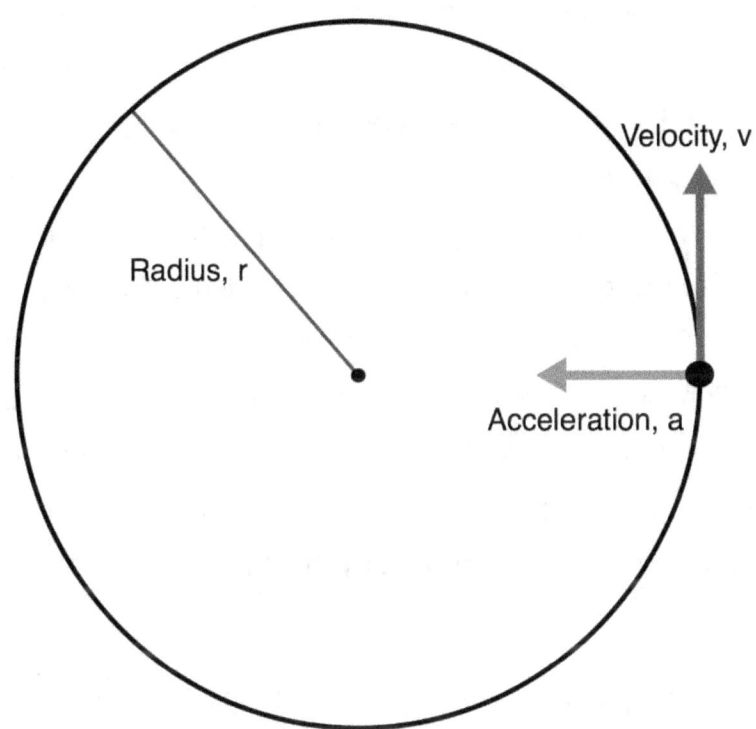

For this singular case, the centripetal acceleration, tangential velocity, and radius of the circle are all connected by the following equation. The magnitude of the acceleration is equal to the square of the object's linear speed divided by the radius of the circle.

$$Centripetal\ Acceleration = \frac{linear\ speed^2}{radius}$$

$$a = \frac{v^2}{r}$$

Additionally, the object's linear speed and angular speed are connected by the circle's radius.

$$Linear\ Speed = radius \cdot angular\ speed$$

$$v = r\omega$$

Because the object is rotating at a constant rate, it has zero angular acceleration.

We can also easily determine the time it takes an object to make one full rotation. This is also called the **period**, T. Essentially, we are using our standard velocity equation, setting the distance to be the circumference of the circle, and then solving for the time. The time to complete one full circle, the period, is then equal to two times pi times the radius of the circle, divided by the linear velocity of the object.

$$Time\ to\ Complete\ One\ Circle = Period = \frac{2\pi \cdot radius}{linear\ speed}$$

$$T = \frac{2\pi r}{v}$$

Let's use these equations to figure out how fast Earth orbits the sun, as well as its centripetal acceleration. The period of orbit is one year, which is roughly equal to 31.6 million seconds. On average, Earth is roughly 93 million miles or 150 billion meters away from the sun. We can rearrange the equation for the period to solve for Earth's speed as it orbits! We will also use scientific notation extensively to remove some clutter.

$$v = \frac{2\pi r}{T} = \frac{2 \cdot 3.14 \cdot (150 \cdot 10^9 \, m)}{31.6 \cdot 10^6 \, s} = 30{,}000 \, \frac{m}{s}$$

If we want the angular speed, we can divide the linear speed by the radius of the circle.

$$\omega = \frac{v}{r} = \frac{30{,}000 \, \frac{m}{s}}{150 \cdot 10^9 \, m} = 2 \cdot 10^{-7} \, \frac{rad}{s}$$

Lastly, let's use that linear speed to solve for Earth's centripetal acceleration.

$$a = \frac{v^2}{r} = \frac{\left(30{,}000 \, \frac{m}{s}\right)^2}{150 \cdot 10^9 \, m} = 6 \cdot 10^{-3} \, \frac{m}{s^2} = 0.006 \, \frac{m}{s^2}$$

In the next chapter, we'll look at Forces, which are what cause objects to accelerate.

Ch. 2 – Mass, Momentum, and Forces

New Motion Terms

In the last chapter, we introduced and formalized the basic terms to describe motion: position, displacement, velocity, and acceleration. Now, we will look at the causes for acceleration, which are **Forces**, F. Forces can be thought as pushes or pulls on an object. In the case of gravity and electromagnetism, the objects do not necessarily need to be in direct contact for a force to occur. Forces are vectors, so we have to be mindful of direction. If you are to push an object, you are always pushing it in a certain direction. With Forces, we have our first complex unit with its own unique name and symbol. The base metric units for Forces are Newtons, N, which are the same thing as kilograms, kg, times meters per second squared (kg•m/s^2).

We know that it takes much more force to move heavier objects than lighter objects. **Mass**, m, is an object's resistance to Forces. The base metric units for mass are kilograms, kg. Mass is an intrinsic quantity for an object, as mass is a statement about the amount of matter in that object. **Weight**, however, is the force of gravity on something, and is a derived quantity. An object's *weight* changes on the moon, but its *mass*, the amount of matter in it, is always the same.

With mass, we can also define another key term for motion, which is **momentum**, p. Linear momentum is equal to an object's mass times its velocity. Its units are just kilograms times meters per second, kg•m/s.

$$Linear\ Momentum = Mass \cdot Velocity$$

$$p = mv$$

Like velocity, momentum is a vector, so direction is part of it. With momentum, we can, for example, compare the relative effects of a fast-moving car to a slow-moving train.

To give some numbers, let's have an average sized car of 1400 kg moving at a respectable 20 m/s. Its momentum is therefore 28,000 kg•m/s. A train has a much larger mass of 50,000 kg but is only moving at 5 m/s. Its momentum is then 250,000 kg•m/s – far larger due to its much higher mass.

Newton's Laws

Mass, Momentum, and Force are all connected through Newton's Three Laws of Motion. Newton's Laws describe how Forces function and give them mathematical meaning. There are a few ways that Newton's Laws can be written, so you can expect to see some variation in their descriptions.

Newton's First Law of Motion:

Objects at rest remain at rest, and objects at a constant velocity remain at that velocity, unless acted on by a net force.

More precisely, a system's* total momentum is constant unless acted on by a net external force.

The first part of the first law is fairly straightforward – objects that aren't moving will stay that way until something is done to them. If you place an object on the floor, it's going to stay there until you or someone or something pushes it.

The second part is much harder to see in everyday life. This is because gravity and friction are always present around us. If you throw an object, giving it some initial velocity, it will fall to Earth. Likewise, if you push something across the floor, it will eventually come to a stop due to friction. This is a case where we have to look at an ideal scenario – a common tactic in physics, as real life tends to be very complicated! You can picture moving an object on a cart with well-oiled wheels. Once you give it a push, it will

---

* A system refers to a fixed number of objects or particles

continue along at that velocity for some time. Similarly, you can push something across a frozen lake or ice rink without it slowing down. In those cases, friction is greatly minimized, so the object's velocity is roughly constant.

You can also apply this first law to an object's momentum. If all the forces are internal (i.e., within the system or the object), then the total momentum of an object or system is constant. Imagine a firecracker exploding in the sky. Parts of the firecracker scatter in all directions. Each piece that goes left, for example, has a certain amount of leftward momentum. That will be counterbalanced by other pieces of the firecracker that go right, with their having rightward momentum. This is because the force of the explosion was all inside the firecracker system and nothing from the outside was acting on the firecracker (aside from gravity and a little bit of air resistance). We'll use this principle to examine collisions later in this chapter.

Newton's Second Law of Motion:

The sum of the forces acting on an object is equal to its mass times acceleration.

$$\Sigma F = ma$$

More precisely, the sum of the external forces acting on an object is equal to its change in momentum divided by its change in time.

$$\Sigma F = \frac{\Delta p}{\Delta t}$$

In the above equations, the symbol, Σ, is the uppercase Greek letter sigma, and means "sum of" for that type of variable, so ΣF means "sum of forces."

This second form of Newton's Second Law is the most general. However, in most cases, the object's mass is constant, so only the velocity changes, which then results in the object's change in momentum over change in time just equaling its mass times acceleration.

It is through the Newton's Second Law that we get the mathematical meaning of forces. Forces are what cause objects to accelerate or to change their linear momentum. When physicists use forces to examine a particular scenario, we are basically looking to solve this equation. We try to identify all the forces present, being mindful of their direction, and then solve for the acceleration of the object. We'll do some examples of this once we identify some specific forces.

You can derive Newton's First Law from this Second Law. When no external forces are present, there is no acceleration or change in momentum, so therefore the momentum of the system is constant. As soon as net force is present, acceleration occurs and the momentum changes.

Again, in the fireworks case, all the forces are internal, so no external forces mean that the total momentum of the firecracker system is constant.

<u>Newton's Third Law of Motion:</u>

For every force or action, there is an equal and opposite force or reaction.

This is basically a statement about how nothing happens in isolation. Forces always come in pairs. Just as the Earth exerts a gravitational force on the moon, the moon exerts a gravitational force on Earth. If you push a wall, that wall pushes back on you. It is because of friction between you and the floor that keeps you in place. If you were wearing skates or doing so on ice, you would move backward due to that push back.

Newton's Third Law does not mean that everything cancels out. Each object can have different sets of force pairs acting on it. When we analyze systems using forces, we are normally just interested in a single object. For example, when we look at pushing a sofa across the floor, we're not worried about all of the forces acting on the floor, the Earth, and the pusher – only what is happening to the sofa. However, the forces that the sofa exerts on the objects around it may need to be considered to account for the push backs on the sofa due to Newton's Third Law.

A subtle point that can be inferred from Newton's Third Law is that when you look at two objects interacting, it should not matter whether you observe the forces from object A's perspective or object B's perspective. The forces acting between them are not unique to A or to B – they exist for both. In other words, the Laws of Physics should be the same regardless of whether you have Object A's perspective or Object B's perspective.

Specific Forces

We'll now turn our attention to some specific forces and some case studies in how to use those forces and Newton's Laws to analyze the motion of various objects and systems.

Full form of Gravity

Another of Newton's key discoveries was the Law of Gravity – the equation that describes how the Sun interacts with the planets, and how Earth interacts with the moon and yourself, and so forth. The force of gravity between two masses, M and m, is determined by the value of those masses and the distance between them, r.

Full form of Gravity

$$F = \frac{GMm}{r^2}$$

Here, G is the gravitational constant and is equal to 6.67 •10⁻¹¹ N•m²/kg². The direction of the force is always attractive or towards each other.

If you look at the force between the Earth and Sun, it doesn't matter which one you call "m" and which one you call "M." This is an example of Newton's Third Law in action. The force of gravity between the Sun and the Earth *has* to be the same for each object.

The term, tidal force, refers to how gravity affects closer objects more strongly than further objects. As a result, objects become stretched out. We see this in Earth's oceans. High tides occur when the moon is closest to that part of Earth or furthest away. When the moon is closest, the oceans are drawn more closely toward it, so the water is pulled further up. When the moon is furthest away, the water is not pulled very strongly from the moon, so it is freer to pull away from the Earth.

It turns out that unless you are dealing with something moon-sized or greater, the force of gravity is extremely small. This is because that gravitational constant, G, is so tiny. For example, let us look at the force between two basketballs that are right next to each other. The mass of a basketball is about 0.6 kg, and let's place them 20 cm apart from each other. Solving for the gravitational force between them, we have the following.

$$F_{Ball\ on\ Ball} = \frac{GM_{Ball}m_{Ball}}{r^2_{Ball\ to\ Ball}}$$

$$= \frac{\left(6.67 \cdot 10^{-11}\ N \cdot \frac{m^2}{kg^2}\right) \cdot (0.6\ kg) \cdot (0.6\ kg)}{(0.2\ m)^2}$$

$$F_{Ball\ on\ Ball} = 6.003 \cdot 10^{-10}\ N$$

This is an awfully small amount. It turns out that the planet Saturn has a much bigger effect on the basketball than a second basketball located right next to it. At its closest, Saturn is about 746 million miles ($1.2 \cdot 10^{12}$ m) away from Earth and has a mass of $5.7 \cdot 10^{26}$ kg. Solving for its gravitational force, we have the following.

$$F_{Saturn\ on\ Ball} = \frac{GM_{Saturn}m_{Ball}}{r^2_{Ball\ to\ Saturn}}$$

$$= \frac{\left(6.67 \cdot 10^{-11}\ N \cdot \frac{m^2}{kg^2}\right) \cdot (5.7 \cdot 10^{26}\ kg) \cdot (0.6\ kg)}{(1.2 \cdot 10^{12}\ m)^2}$$

$$F_{Saturn\ on\ Ball} = 1.58 \cdot 10^{-8}\ N$$

This is still a small amount, but it's over twenty times greater than the effect of that nearby basketball! This means that the other planets of the solar system have a greater gravitational effect than all of the objects around you!

Gravity near surface of Earth

If we look at the gravitational effect of Earth on objects on the surface, we can surely use the full form of gravity to do so. Here, we would use the mass of Earth, $5.972 \cdot 10^{24}$ kg, and its radius, 6,356 km to 6,378 km. Because those distances are so great, there's no appreciable difference in the force of gravity from Earth if you add or subtract a hundred meters in the process. We can therefore solve for the acceleration due to Earth's gravity, g, by using the full form of gravity, removing the second mass, m, and using its average radius:

$$Acceleration\ due\ to\ Earth's\ Gravity$$

$$g = \frac{GM_{Earth}}{r_{Earth}^2}$$

$$= \frac{\left(6.67 \cdot 10^{-11}\ N \cdot \frac{m^2}{kg^2}\right) \cdot (5.972 \cdot 10^{24}\ kg)}{(6.36 \cdot 10^6\ m)^2}$$

$$g = 9.8 \frac{m}{s^2}$$

Reinserting that value for the acceleration due to gravity into the full form of gravity equation, we have the following simple equation for the force of gravity near Earth's surface on an object with a mass, m.

$$Force\ of\ Gravity\ near\ the\ Surface\ of\ Earth$$

$$F = mg$$

Again, g is the acceleration due to Earth's gravity and is equal to 9.8 m/s^2.

An important result from the above derivation is that although the *force* of gravity is based on the mass of the object, the *acceleration* due to gravity is the same for all objects. Therefore, if friction or air resistance is not a factor and gravity is the only force present, all objects fall at the same rate (9.8 m/s^2), regardless of their mass! The story of Galileo dropping various objects off of the tower of Pisa and their landing at the same time reflects that.

Normal Forces

One important result of Newton's Third Law is the Normal Force, which is the "push back" between objects in direct contact. The reason for the name is that the direction of the force is Normal or Perpendicular to the surface of contact. If you push on the wall, the wall pushing back on you is a Normal Force. The most common example of a Normal Force is due to an object being pressed onto the ground due to gravity. The ground pushing back on the object is a Normal Force. In the case where an object is just staying on the floor and not crashing through it, the force of gravity and the normal force fully cancel each other, so that the net vertical force on the object equals zero. In the case of someone or something pulling up or pushing down on the object, those forces would also be included in determining the Normal Force.

Friction

Friction is something that we encounter all the time. Frictional forces always oppose the direction of motion for a moving object or the direction of "intended" motion for a stationary object. Unfortunately, the mathematics for describing friction can be quite difficult, as it has to do with the collisions between surfaces and molecules. Some of the forms of friction can be readily discussed, however.

First, let us consider the interaction between two solid surfaces. The direction of the frictional force here is always parallel to the surfaces of contact. Although the surfaces may appear flat to the naked eye, they are full of peaks and valleys in the microscopic level, and those rough surfaces collide against each other, causing that friction.

You may have noticed that is very difficult to cause heavy objects to move, but once they are moving, it is relatively easier to keep them going. This is because the force of friction between solid objects changes based on whether motion is occurring or not. In the motionless or *static* case, the force of friction will balance out any other external forces so as to prevent motion. The maximum amount of static friction is based on the normal force acting on the object as well as a unitless constant based on the two materials. Once that maximum amount of static friction is overcome, the object will start to move. At this point, the object is subject to *kinetic* friction. It too is based on the normal force and a constant of the two materials. However, the constant for kinetic friction is always less than that for static friction, which is why it is easier to keep objects moving once they have started. Additionally, because the force of friction between solid objects is also based on the normal force, pressing the object against the solid surface of contact increases the friction. This is why planting your feet helps slow you down.

    Air resistance, or more generally the drag through a fluid, is trickier as it depends on several factors, most notably the velocity of the object and its shape. The faster an object is moving, the greater the air resistance. If an object is dropped from a great height, it will have enough time to reach a terminal velocity, where the force of friction is equal and opposite to the force of gravity acting on the object. Since the net force is now equal to zero, the object experiences no further acceleration, and its velocity remains constant.

    Let us look at a useful case study: moving a sofa across the floor. Again, you have to push a heavy object hard to get it to move so as to overcome the static friction, but once it is moving, less force is required to keep it going.

What forces are acting on the sofa while you are pushing it around? If you are pushing the sofa to the right, friction between it and the floor will act against that motion causing a force pointing to the left. As long as the sofa is on Earth, gravity is acting on it, pushing it down to the floor. Presuming that the sofa is not crashing through the floor, the floor pushes upward with a normal force due to Newton's Third Law. The figure below illustrates those four forces and their directions.

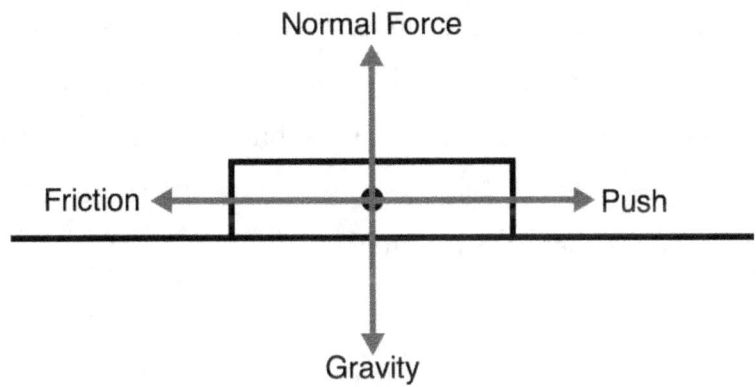

Let's say that the mass of the sofa is 120 kg, and you push it with a 400 N force, and the force of friction in this case is 380 N. What value is the normal force from the floor on the sofa, and what is the sofa's acceleration? Because forces are vectors, we have to be mindful of direction, and treat the horizontal and vertical forces separately.

Again, as long as the sofa is not crashing through the floor, there is no net vertical motion for the sofa. This means that the force of gravity on the sofa must equal the normal force acting on the sofa. Since the force of gravity from Earth is equal to mg, the force of gravity and the normal force are then equal to 120 kg times 9.8 m/s², which results in 1,176 N. If you were pushing *down* on the sofa instead of across, that downward force would also contribute to the normal force that would counteract it.

As for the horizontal direction, you push the sofa to the right with 400 N, and the force of friction opposes this by going left at 380 N. Therefore, the net horizontal force is equal to 20 N. We know from Newton's Second Law that the net force acting on an object is equal to its mass times acceleration ($\Sigma F = ma$). If we divide that net 20 N net force by the sofa's mass of 120 kg, we would get its horizontal acceleration, which is 0.167 m/s².

Restorative Forces

Restorative forces are important in that they show up in a wide variety of contexts. These are forces that act to return a system to its original position or configuration. The most common example of a restorative force that you encounter in everyday life is for springs. Here, the spring's force is equal to the negative of the spring constant, k, times the displacement, Δx, from its equilibrium position. The spring constant is based on what the spring is made of and how it is coiled. The reason for the negative sign is so that the spring's force is restorative and opposes any displacement from equilibrium.

$$F_{Spring} = -k\Delta x$$

This equation effectively models the motion of atoms in molecules to the overall stress and strains of a material. When you push on the wall, you are effectively displacing the surface atoms by a microscopic amount, and as soon as you stop pushing, the surface returns to its original position according to that spring force.

If you retract or compress a spring, you'll notice that the spring oscillates back and forth in a pattern. The time it takes to make a full cycle is based on that spring constant, the spring's mass, and any other mass attached to it. However, that spring will eventually come to a stop according to some damping or frictional force. That damping force is based on another constant and the velocity of the spring. As time goes on, this damping reduces the maximum oscillation of the spring until it comes to a complete stop.

Collisions

So far, momentum has not been so important in our discussion about forces. However, it is essential when we examine what happens in collisions. The more precise form of Newton's Second Law is that the sum of the *external* forces acting on a system is equal to its change in total momentum over the change in time. In a collision, all the forces are internal, so the combined momentum of the two interacting objects must be constant. You can picture the collision taking place inside an imaginary box. As long as nothing from outside the box affects the inside, the total momentum inside the box is constant.

This can also be inferred from Newton's Third Law. We have two objects, A and B, that collide. We know that the force of the collision on B is equal and opposite to the force of the collision on A. Additionally, the duration of that collision force has to be the same for each object. Therefore, if the times are the same and the forces are equal and opposite, then the change in momentum has to be equal and opposite for the two objects! In a collision, momentum is exchanged between the colliding objects, but the total amount is constant.

In equation form, we would write that initial momentum for A plus the initial momentum for B is equal to the final momentum of A plus the final momentum for B.

Total Momentum of A and B is constant in a Collision

$$Initial\ p_A + Initial\ p_B = Final\ p_A + Final\ p_B$$

Throughout this process, it is important to note that momentum is a vector, so we must be mindful of direction when we do any sort of math.

Let's take a look at few examples. First, we have object A with a mass of 5 kg is moving to the right at 20 m/s, and collides with object B with a mass of 20 kg and is at rest. After the collision, object B is seen to move to the right at 8 m/s, and object A ricochets to the left. How fast is object A moving? The figure below illustrates what is happening.

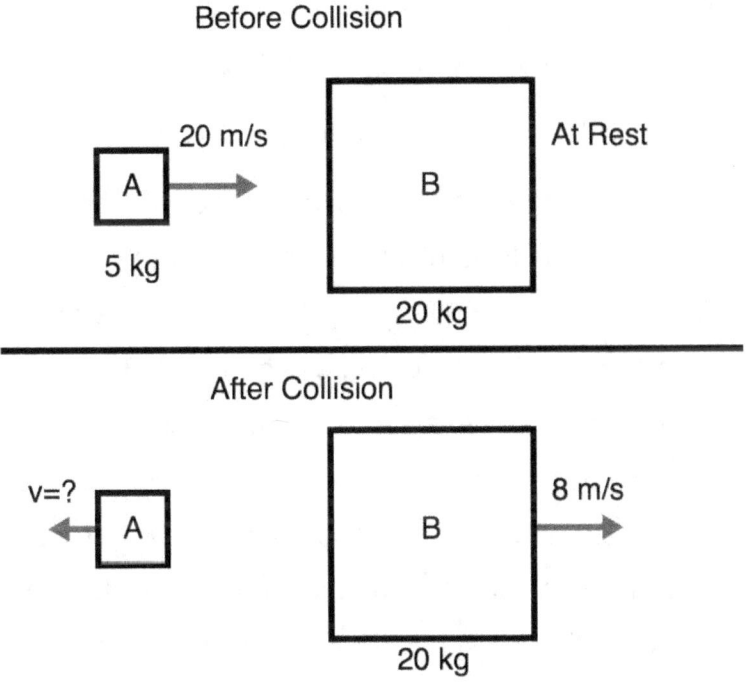

Momentum is equal to mass times velocity and is a vector (p = mv). We'll say that objects moving right have a positive velocity and momentum, and objects moving left have a negative velocity and momentum. The initial momentum of A is its mass, 5 kg, times its velocity, +20 m/s, so the initial momentum for A is +100 kg•m/s. Since B is initially at rest, it has zero initial velocity and therefore zero initial momentum. Therefore, the total momentum of the system is +100 kg•m/s. The final momentum of B is its mass, 20 kg, times its final velocity, +8 m/s, so the final momentum for B is +160 kg•m/s. Let's put these values into our total momentum equation.

$$\text{Initial } p_A + \text{Initial } p_B = \text{Final } p_A + \text{Final } p_B$$

$$+100 \; kg \cdot \frac{m}{s} + 0 = \text{Final } p_A + 160 \; kg \cdot \frac{m}{s}$$

Rearranging terms, we have that the final momentum of A is -60 kg•m/s. If we divide that by A's mass of 5 kg, we have that A's final velocity is -12 m/s! Here, the negative sign indicates that the object is moving to the left.

We can also have a scenario where the objects stick together after the collision. Instead of the blocks being made out of metal, perhaps that are made out of clay. We'll keep the masses and initial velocities the same, but instead of A ricocheting, it sticks to B and they both move forward at some unknown velocity which we'll solve. The figure below illustrates this set-up.

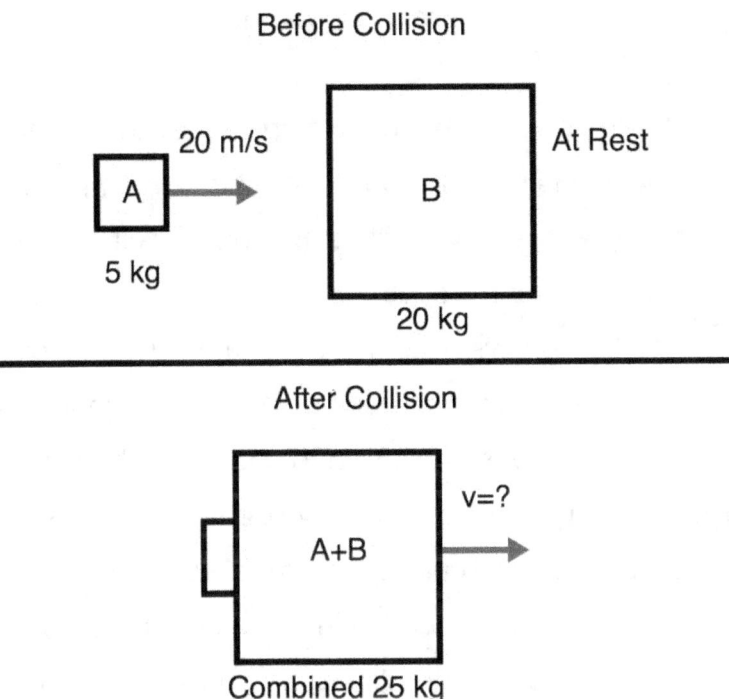

 The initial momentum for each object is still the same, so the total momentum of the system is still +100 kg•m/s. Since the objects stick together after the collision, they function as one object with a combined mass of 25 kg moving at the same final velocity, resulting in the following equation for their total momentum:

Objects stuck together: $Total\ p = (m_A + m_B)v_f$

We can divide the total momentum of the system by that combined mass to get their final velocity of +4 m/s.

Forces in Rotation

Newton's Laws and the examples provided so far have all been for linear motion. Mass, Momentum, and Force can be applied to rotational motion as well. However, each of those terms have to be expanded in the process.

Rotation always occurs about some axis, be it through the objects center, like the Earth, or along an edge, like a door. In order to cause an object to rotate, you have to exert a force in a particular direction. For example, you push or pull on a door perpendicular to the door itself in order to open it. If you push the short side of the door towards the hinge, nothing happens. Additionally, where you push or pull makes a difference. It is easier to push the door on its edge versus near its hinge. The further away your push is from the axis of rotation, the greater its effect.

In short, the cause of a rotation is due to the force, its direction, and the distance between the force and the axis of rotation. The product of these quantities is **Torque**. Its symbol is the lower-case Greek letter tau, τ, and it has units of Newton-Meters, N•m*. Torque has the greatest value when the force is applied perpendicularly to the object.

---

* N•m are also the same units for energy. However, torque by itself is not a form of energy. The rotational energy from a torque is equal to the torque times the angle, in radians, that it is acted over.

$$Torque = Force \cdot Distance\ from\ Axis\ of\ Rotation$$

An object's resistance to torque is based not only on its mass but also the distribution of that mass. This resistance to rotation for an object is called its **Moment of Inertia**, I, which has units of kilograms times meters squared, kg•m². When an object's mass is more spread out or on edge, like for hollow objects and loops, the moment of inertia is high. When the mass is closer to the axis of rotation, the moment of inertia is low.

Linear momentum is equal to an object's mass times its linear velocity. We have a similar relation for rotation, where we swap in rotation's version for each of those quantities. An object's **Angular Momentum**, L, is equal to its Moment of Inertia, I, times its Angular Velocity, ω. The units for Angular Momentum are kg•m²/s.

With these quantities, we can reframe Newton's Laws of Motion for Rotation. Newton's First Law for Rotation is that a system's total angular momentum is constant when no external torques are present. A good example of how this can be applied is for a figure skater. When a figure skater has her arms and legs spread out, she spins rather slowly. She then decides to tuck in her arms and legs. Since no one is pushing on her, no external torques are present, and her angular momentum is constant. By bringing more of her body towards her center, she decreased her moment of inertia. Because her angular momentum must be constant and the moment of inertia has decreased, this means that her angular velocity must increase, so she spins faster!

Newton's Second Law for Rotation is that the sum of the external torques acting on an object is equal to its change in angular momentum divided by its change in time.

$$\Sigma \tau = \frac{\Delta L}{\Delta t}$$

Lastly, Newton's Third Law is that for every torque there is an equal and opposite torque. One important application of this is in helicopters. Helicopters always have two sets of rotors and spinning blades – one set for each action. Otherwise, the helicopter body would be forced to spin! This is also a result of the helicopter's total angular momentum being constant due to the lack of any *external* torque.

In the above descriptions, we only looked at rotation along one axis. You can have rotation along up to three axes (for example, imagine a physics book tumbling through the air). As soon as there is rotation in multiple directions, things get very messy and involve matrix math. Even in the linear force examples, we limited ourselves to perfectly horizontal or vertical forces. This is because vector manipulation can be difficult. However, we can avoid some of these issues by doing a scalar treatment of motion – Energy, which is the subject of next chapter.

Ch. 3 – Energy

Energy in Mechanics

So far, nearly all of the terms and variables we have used to describe motion and its causes have been vectors. Through a mathematical process called integration, we can reframe Newton's Second Law in terms of scalars. Not only are scalars mathematically simpler, the scalar approach ends up being more general. These new scalars are going to be forms of **Energy**.

Energy is an extremely important concept in Physics, as we use it in just about everything. Unfortunately, it can also be somewhat abstract. For now, we'll define Energy as being a scalar quantity that describes a system's capacity to exert forces over distances. The general symbol for Energy is E, and its units are Joules, J, which are the same as kg•m^2/s^2 or N•m. Energy comes in a variety of forms, some of which will be introduced in future chapters. For now, let's consider the main forms of Energy in mechanics.

Kinetic Energy

Our first form of energy is **Kinetic Energy**, K or T, which is the energy associated from an object's motion. Fortunately, we have exact, simple equations for it. For linear motion, the linear kinetic energy is equal to one-half of the object's mass, m, times its linear speed, v, squared[*].

$$Linear\ Kinetic\ Energy = \frac{1}{2} \cdot mass \cdot linear\ speed^2$$

$$Linear\ K = \frac{1}{2}mv^2$$

There is also rotational kinetic energy, which is one-half of the object's moment of inertia, I, times its angular speed, ω, squared[†].

Rotational Kinetic Energy

$$= \frac{1}{2} \cdot moment\ of\ inertia \cdot angular\ speed^2$$

$$Rotational\ K = \frac{1}{2}I\omega^2$$

For much of this chapter, we'll ignore rotational motion in order to simplify things.

---

[*] You can also express the linear kinetic energy in terms of linear momentum. This becomes very important in quantum mechanics. $Linear\ K = \frac{p^2}{2m}$

[†] Likewise, the rotational kinetic energy can be expressed in terms of angular momentum: $Rotational\ K = \frac{L^2}{2I}$

Potential Energy

Our next form of Energy is **Potential Energy**, V or U, which is the energy associated with forces and an object's position or configuration. Potential energy is sometimes referred to as a system's "stored" energy. Because it is based on the forces present, there are wide variety of potential energies[*]. However, there is a very important consequence of potential energy in all its forms – if possible, objects and systems move and evolve to reduce their potential energies. To demonstrate this, let's look at the potential energy from Earth's gravity near its surface. It is equal to the object's mass, m, times the gravitational acceleration from Earth, g, times the relative height of the object, h. Here, the relative height typically means the height above ground level.

$$V_{Earth} = mgh$$

If you hold an object up high in the air, you know that it will fall to the ground due to Earth's gravity. Last chapter, we talked about that in terms of the force of gravity, mg. You can also talk about the object falling in terms of its potential energy. If you hold an object up high, it has a certain amount of potential energy due to that height. As soon as

---

[*] For Forces that are constant or dependent exclusively on position, the Potential Energy is the negative of the integral of the Force times the displacement

you let go, the object falls, thus reducing its height and its potential energy. This same sort of thing occurs for *any* physical system. Objects move and rearrange themselves to reduce their potential energies. Chemical reactions occur to reduce the potential energy of the interacting atoms.

Last chapter, we introduced several specific forces. We can reframe them in terms of energy. For the full form of gravity, the potential energy takes on the following form.

$$V_{Gravity} = -\frac{GMm}{r}$$

As objects get closer to each other, the distance between them, r, is reduced. As r gets smaller, 1/r gets larger. Since the potential energy has that negative sign, the potential energy gets more and more negative (i.e., smaller) as the two objects approach each other.

An important form of potential energy is that for restorative forces, such as occur in springs, material deformations, and interatomic bonding. Its potential energy is as follows.

$$V_{Springs} = \frac{1}{2}k(\Delta x)^2$$

As a spring is stretched or compressed, the displacement from equilibrium, Δx, gets larger. As soon as you let go, the spring tries to return to its original configuration, thereby reducing its displacement from equilibrium as well as its potential energy.

We also talked about normal forces, which were the "push back" between objects in direct contact according to Newton's Third Law. It turns out that there is no energy associated with normal forces since they do not directly contribute to an object's motion. This gives us one less thing to worry about!

Heat

Unfortunately, the energy from frictional forces are not considered forms of potential energy. This is because the effect of friction is very dependent on the path that is taken and the amount of time it is present. However, we still need to account for friction's energy in terms of heat. Heat is a major component of thermodynamics – the subject of a later chapter.

For now, we'll define **Heat**, Q, as the energy associated with a change in temperature as a result of friction. Since this energy is dependent on the path and time that friction occurs, there is no simple equation to define it in those terms.

The Conservation of Energy

The main reason why we care about Energy in Physics is that it is a *conserved* quantity. **The Conservation**

of Energy means that the total energy of a system is always constant, no matter what. Alternatively, energy cannot be created or destroyed, only transferred from one form or system to another. In every field of physics, we look to apply the conservation of energy in some fashion. We find what the total energy is, and look at how that energy gets distributed in different circumstances. If we are able to find the total energy at one moment, then we know the total energy at all times and locations.

For the forms of energy we have discussed so far, the total energy of a system would be its kinetic energy, potential energy, and heat (if there's friction) all summed together.

$$Total\ Energy = Kinetic\ Energy + Potential\ Energy + Heat$$
$$= constant$$
$$Total\ E = K + V + Q = constant$$

Let's look at a few examples. You decide to push a block across a rough surface. That block has some initial kinetic energy due to that push. As the block slides across that rough surface, it slows down due to friction, and its kinetic energy gets transferred to heat.

If you drop an object from on high, it has a certain amount of potential energy due to Earth's gravity. As it falls, its potential energy decreases, and that energy gets transferred to kinetic as it speeds up, and a little bit of energy goes to heat due to air resistance.

For a spring's motion, the initial stretching or compressing a spring results in that spring having a certain amount of potential energy. Once you let go, that spring starts to move and gains some amount of kinetic energy. As the spring oscillates, its energy constantly changes between potential and kinetic. If spring damping is included, some of that energy converts to heat each cycle.

The ease of using energy to analyze systems really shines in cases where friction is minimized or can be ignored. Let's look at a block sliding down a smooth, frictionless incline plane, as shown in the figure below. The incline is at a 30° angle. The block has a mass of 6 kg and starts at rest at the top. The incline is 20 m long and 10 m high. How fast is the block moving when it reaches the bottom?

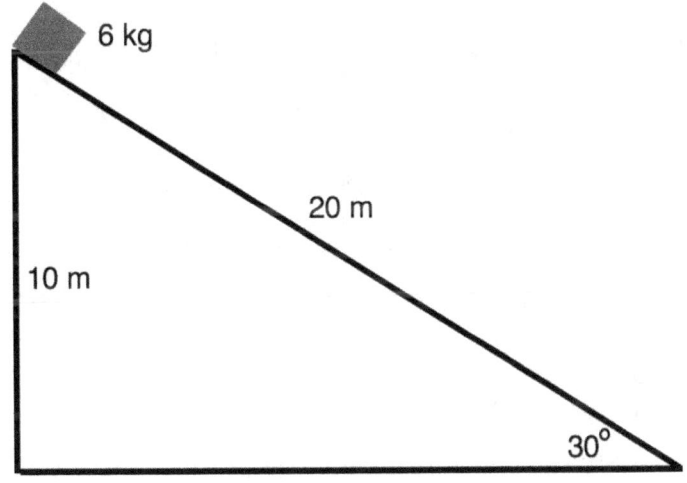

You can solve this problem using our motion equations and Newton's Laws. However, it is messy and involves some trigonometry. The way we use energy to analyze what is happening is to find the total energy of the system and see how that energy gets distributed at the bottom. We know everything about what is taking place at the top of the incline, so let's start there.

First, the block starts at rest, so it has no kinetic energy. Even if the incline were rough, we can always set the initial heat to equal zero, since no energy has been transferred to heat yet. This is similar to how we can usually set the initial position or initial time to equal zero. All that we have is the block's potential energy. We have its mass of 6 kg, g is a constant 9.8 m/s², and it starts at a relative height of 10 m above the bottom of the incline – only the vertical distance is important here. We multiply those three quantities to get the block's initial potential energy.

$$V_{Earth} = mgh = 6 \, kg \cdot 9.8 \frac{m}{s^2} \cdot 10 \, m = 588 \, J$$

The total energy of the block at the top is its kinetic energy (zero, since it's at rest) plus its potential energy (588 J) plus its heat (always zero initially), giving a total energy of 588 J.

Due to the conservation of energy, the block's total energy has to be 588 J at the bottom as well. Let's find each form of energy. First, since we made the incline frictionless, no energy gets converted to heat. Second, since the block is at the bottom, its relative height above ground is zero, so it has no final potential energy. This means that all 588 J goes to kinetic energy at the bottom!

Since we have that final kinetic energy, we just need to rearrange the kinetic energy equation to solve for the speed at the bottom, which ends up being 14 m/s.

$$K = \frac{1}{2}mv^2 \rightarrow v = \sqrt{\frac{2K}{m}}$$

$$v = \sqrt{\frac{2K}{m}} = \sqrt{\frac{2 \cdot 588\,J}{6\,kg}} = 14\,\frac{m}{s}$$

There are two other terms we can introduce regarding the conservation of energy and how energy transfers. First, let's go back to pushing a block across a rough surface. Initially, that block was at rest until you gave it a push. It started with zero kinetic energy, and, after you pushed it, it now had some amount of it. This is due to the **Work** that you did to the block.

Work, W, is defined as a force, F, applied over a displacement, Δx. As a result of that Work, an object gains a certain amount of Kinetic Energy, ΔK. Work is one way energy can transfer from one object or system to another*.

---

* You can apply work to rotational motion as well. In this case, the rotational work is equal to a torque applied over an

$$Work = Force \cdot Displacement = Change\ in\ Kinetic\ Energy$$
$$W = F\Delta x = \Delta K$$

You may have noticed that *time* has not shown up in any of these equations. If you want to look at the time behavior of these energy exchanges, you use the term **Power**. Power, P, is the rate that energy changes over time. Remember, the total energy of a system is always constant. You use Power to look at how a particular *form* of energy changes in time (change in kinetic energy over time, for example). The units for Power are Watts, W, which are the same as J/s.

$$Power = \frac{Change\ in\ Energy}{Change\ in\ Time}$$

$$P = \frac{\Delta E}{\Delta t}$$

Simple Machines

An important application of all the concepts that we have discussed so far is the six simple machines: the incline plane, screw, wedge, lever, pulley, and wheel and axle. Each machine either amplifies and/or redirects forces and employs work. A certain amount of force is applied by the user, called the Input Force, to one part of the machine to

---

angle, which then results in a change in rotational kinetic energy.

overcome some resistance, called the Output Force, to move an object. The ratio of these two quantities is called the **Mechanical Advantage**.

$$Mechanical\ Advantage = \frac{Output\ Force}{Input\ Force}$$

As in all physical systems, the total energy must be conserved. A certain amount of work is put into the system, and a certain amount of work is put out. In any real system, the output work is lessened due to friction, which we'll ignore for the sake of simplicity. The efficiency of a real machine is the ratio of the output work to input work. Work is equal to the force times a displacement. In the context of these machines, what happens is that a small force applied over a large distance results in a large force applied to a small distance. The large force to be overcome is typically gravity acting on a heavy object.

$$Input\ Work = Output\ Work$$
$$Small\ Input\ Force \cdot Large\ Distance$$
$$= Large\ Output\ Force \cdot Small\ Distance$$

Incline Plane

Lifting heavy objects straight up is difficult. It is far easier to push them up a slight ramp or incline plane. The force used to push the object up the ramp is the input force, and the output force (or force to be overcome) is gravity acting on that heavy object.

The downside of using the incline plane is that you have to push the object for a greater distance. In terms of Work, the input work is that force of the push times the distance you move the object. The output work, which again must equal the input work if friction is not a factor, is the force of gravity on the object, mg, times the height that it is lifted*.

Screw

The screw is nothing more than an incline plane wrapped around a cylinder. In the case where you use a screw in construction, the applied or input force is to rotate while pushing the screw. The resistive force is what it would take to push the screw directly into the material, like a nail. Because the screw is slicing into the wood, there is heavy resistance to pulling the screw directly out, which is why screws are used as fasteners. Like for the simple incline plane, the distance for the input force is the length of the

---

*You may notice that the output work equation for the incline plane is the same equation for the gravitational potential energy of an object. This is because that output work is done to overcome gravity's effect on the object.

screw's thread, and the distance for the output force is the length of the screw's shaft.

Wedge

The wedge is two inclined planes connected to each other, as shown in the following figure.

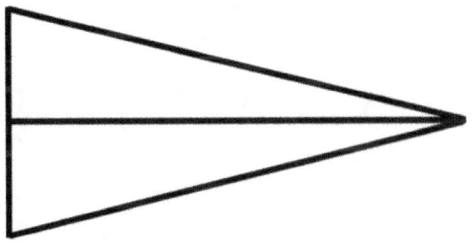

The wedge is used to pierce a material and spread it outward. Like in the screw, the applied or input force goes along the inclines, and the resistive or output force is what it would take to directly split the material a certain width.

Lever

A lever is a plank that can freely rotate about a fulcrum with applied and resistive forces at two different locations. The fulcrum and the forces can be rearranged for different classes of levers, but the physics is unchanged. The figure below is a first-class lever, where the fulcrum is between the two ends of the plank, the input force is at the long end of the plank, and the resistive force is at the short end.

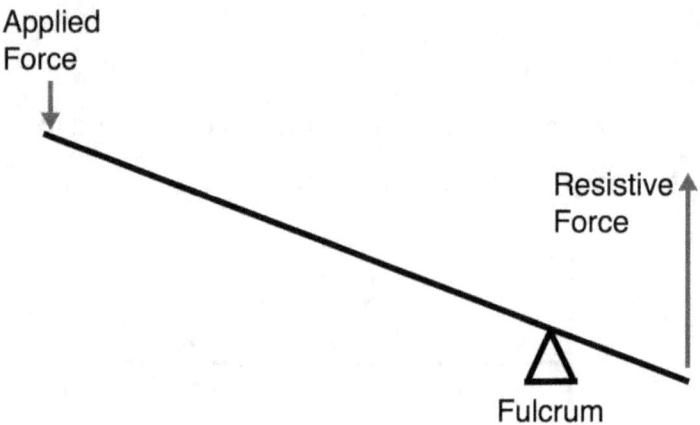

The resistive force is typically gravity acting on an object. If a small force is applied on the far end of the lever, a heavy object can be lifted a short distance on the other end of the lever. Again, this is due to the conservation of energy and the work having to be equal on either end.

Pulley

A single pulley is like a lever with equal lengths on either side of it. It does not increase or decrease the applied force – it just changes its direction. The mechanical advantage comes from using multiple pulleys together in a block and tackle system. The way it works is that you have one set of pulleys fixed to a support and other set of pulleys attached to the object that is to be lifted. The figure below uses a single pulley each, with the rope secured to the top.

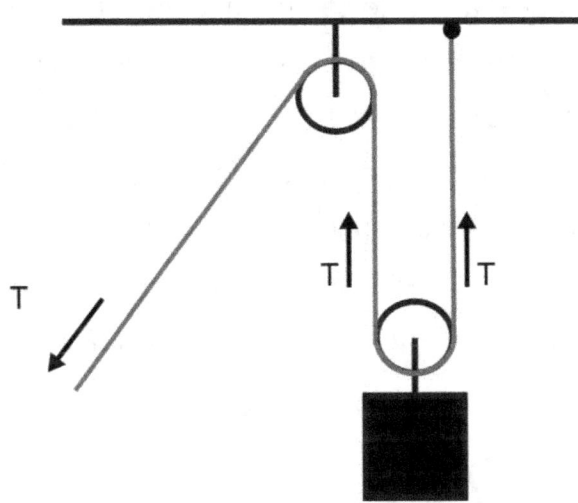

The applied force is your pulling on the rope, giving the rope a certain amount of Tension, T. The resistive force is gravity acting on the lifted object. By having two strands of rope pulling up on the object, the applied force or tension needs to only equal one-half of the weight of the object. The downside is that you now have to pull twice the length of rope than before. This is because the work you put into the system by pulling on the rope must equal the work put out by lifting the object. Adding another set of two pulleys would decrease the needed force by another half but double the length of rope needed to be pulled once more.

Wheel and Axle

The wheel and axle can be thought of as a circular, three-dimensional lever. It is a set of two concentric circles around a single axis that acts as a fulcrum. The forces to cause rotation occur at the edges of their circles, so the distances moved are going to be proportional to their radii. In terms of work, a small force acted on over a larger radius circle will result in a larger force over a small radius circle.

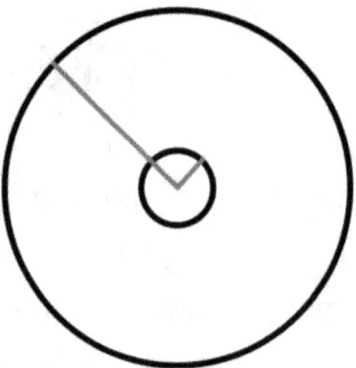

For further reading on simple machines, I would strongly recommend "Basic Machines and How They Work," which is an old training manual produced by the United States Navy that is readily available in paperback or online.

Ch. 4 – Thermodynamics

Macroscopic vs Microscopic Systems

So far, we have looked at the motion of single objects. For the next few chapters, we will be looking at what goes on inside those objects. In Thermodynamics, the study of temperature and heat, we are concerned with the behavior of a very, very large number of particles (say, 10^{20} or more). It is impossible to track the individual motion and forces for each individual particle. Instead, we look at how the group functions on average or as a whole. For this reason, Thermodynamics is also referred to as Statistical Mechanics.

This bird's eye approach to a very large number of moving particles is **macroscopic**. If we instead want to know how each particle individually functions, we would have a **microscopic** approach. Certain variables and terms in Thermodynamics only make sense at the large-view, macroscopic level, and we have to be careful in our language moving forward.

Macroscopic State Variables

To start, we'll examine a few scalar macroscopic variables for describing what goes on inside a substance: temperature, pressure, and density. Because these variables are macroscopic, they only make sense when applied over a large number of particles.

Temperature, T, is traditionally used to describe how hot or cold something is, but those are ultimately subjective terms. What do hot and cold actually *mean* inside a substance? Temperature is a measurement of the kinetic energy of the particles that make up a substance. It is inherently a *local* measurement – when you measure the temperature, you measure it at a particular spot at a particular time. For historical reasons, temperature is not directly measured in units of energy. The base metric units for temperature are Kelvin, K. The reason for this is to connect the baseline of zero kelvin to the fundamental meaning of temperature. Zero kelvin corresponds to the particles having zero kinetic energy. This is known as Absolute Zero. For quantum mechanical reasons, particles cannot reach absolute zero, but can come infinitesimally close to (less that 10^{-10} K has been measured). The temperature in Kelvin is equal to the temperature in Celsius plus 273.15: $T(K) = T(°C) + 273.15$.

For most solids, a change in temperature normally results in a change in dimensions of the object. In most cases, an increase in temperature causes the object to expand, and a decrease in temperature causes it to contract. The relevant equation describing this **linear thermal expansion** is that the object's change in length, ΔL, is equal to the substance's coefficient of linear expansion (a property of the material), α, times the object's original length, L, times the change in temperature, ΔT.

$$\Delta L = \alpha L \, \Delta T$$

The coefficient of linear expansion is a property of the material. It is usually a small positive number (around 10^{-5} per Kelvin). A notable exception is for water – its volume increases by about 9% as it freezes, which is why ice floats!

The main reason why we care about Temperature in Thermodynamics is due to the following property: over time, two systems in direct contact will reach the same temperature throughout. When this happens, we say that they reach **thermal equilibrium**. This is a bit of an idealized scenario, since there are always slight variations in temperature throughout a body, and everything is indirectly in contact with everything else. When dealing with Thermodynamic problems, we are usually interested in what happens at thermal equilibrium, what kind of energy transfers are required, and what this final temperature should be.

Pressure, P, is defined as the Force, F, applied over a certain area, A: $P = F/A$. The units for pressure vary wildly between the sciences and engineering. In Physics, the base metric units for pressure are Pascals, Pa, which are the same as a Newton divided by a meter squared: 1 Pa = 1 N/m^2. Frequently, kilopascals (kPa) are used as well, since atmospheric pressure at sea level is about 101 kPa. In the context of materials, Pressure is used to describe the forces being exerted between the particles inside the substance either from interatomic bonding, in solids, or collisions, in liquids and gasses. In short, Pressure is a measurement of particle interaction. For fluids, pressure is often mistakenly used in place of the fluid's speed. Pressure is a statement of what is going on inside that fluid, and does not necessarily correlate with a high speed.

Density is defined as an object's mass, m, divided by its volume, V. The symbol for density is the lower-case Greek letter rho, ρ, so $\rho = m/V$. Its metric units are kilograms per meter cubed: kg/m^3. Density is a property of a material and its phase. It is largely constant, but can vary slightly with temperature and pressure. This is especially true for gasses. Density is also affected by the substance's purity.

The Three Basic Phases of Matter

There are three general phases of matter (solid, liquids, and gasses). What defines each phase is the amount of interatomic or intermolecular bonding – solids have strong atomic bonds, liquids have moderate atomic bonds, and gasses have little to no atomic bonds. The phase of matter can be determined by the temperature and phase. Solids exist at high pressures and low temperatures, and have high densities. Gasses exist at low pressures and high temperatures, and have low densities. Liquids exist at moderate values for each of those variables. A common graph that is used to describe the phases of a material under varying conditions is a PT diagram (Pressure-Temperature), as shown in the figure below.

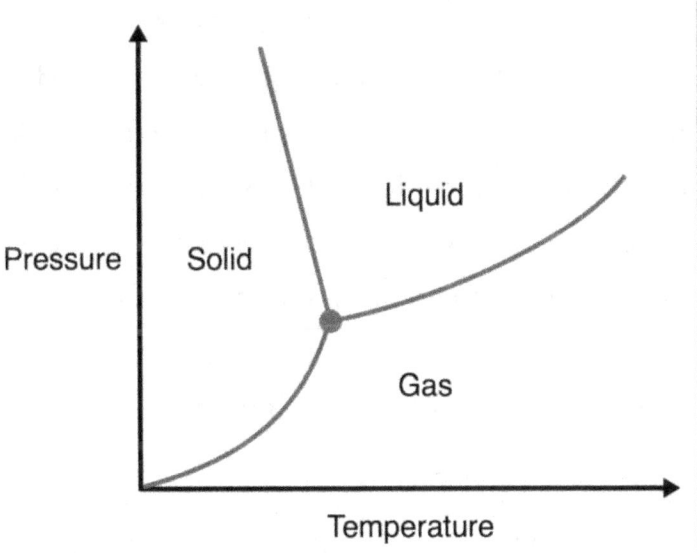

One important feature of this graph is the point where the lines intersect. This is called the **Triple Point** – the specific temperature and pressure where a substance can exist in any of the three phases. For water, the triple point is at 0.01 ºC and roughly 612 Pa.

Solids, having strong atomic bonds, have rigid bodies. They have set volumes, and their surfaces are constant except under strong external forces. The atoms in a solid tend to have fixed locations and only vibrate in place[*]. Those atoms can have a disordered (amorphous) or ordered (crystalline) structure. Crystalline structures come in a wide variety of shapes, from simple cubes to offset hexagons, and much research is done to discover the different possibilities. Depending on the temperature and pressure, a material can exist in several different solid forms.

Liquids have moderate atomic bonds. They also have set volumes, but their surfaces are easily malleable by external forces. In the absence of any other forces, like gravity, they form into perfect spheres so as to minimize their surface tension. Atoms in the liquid continuously interact with each other but can move freely in that volume.

Gasses have little to no atomic bonds, and so their atoms are essentially free to move around. They primarily

---

[*] Atoms on the surface of a solid can change location more easily, which matters a great deal as solids approach the nanoscale.

interact via collision, and will fill whatever volume that is available to them. If the atoms of the gas become ionized, they become plasmas, which have additional electromagnetic properties.

Energy in Thermodynamics

Forms of Energy in Thermodynamics

For historical reasons, Temperature is not measured in units of energy, but it does contribute to the *forms* of energy used in Thermodynamics. In the most basic convention, there are three forms of energy to consider: Work, Internal Energy, and Heat. Work, W, is the same Work used in mechanics – forces applied over a distance. In the context of thermodynamics, it is most often used when an object is compressed or expands. This becomes very important when dealing with gasses.

Internal Energy, U, is a bit of a catch-all form of energy. In this variable, we group together everything that is going on inside the system – how the particles move and how they interact with other. Thus, the internal energy is equal to the total amount of kinetic energy and potential energy for the particles in that system. Since temperature is a measurement of that kinetic energy, the internal energy of a system is proportional to the temperature. For ideal gasses, where the particles are not chemically interacting and therefore have no appreciable potential energy, we can actually have a straightforward equation connecting the internal energy to the temperature and number of gas particles*.

Heat, Q, is the form of energy we need to be careful about. A physics definition would be that Heat is the macroscopic form of energy that is spontaneously transferred between systems at different temperatures. Let's parse this out a bit. First, heat is macroscopic. We are not immediately concerned with how one individual particle exchanges energy with another individual particle – we are just looking at the net effect of a very large number of exchanges. For all intents and purposes, Heat is all of the energy exchanges that cannot be considered Work. Second, heat is a flow or a process. We can talk about heat flowing from A to B or B to A, but it is not accurate to say that A has more or less heat than B or it holds on to heat. We can only compare their temperatures and internal energies in this way. Due to the conservation of energy, the total heat leaving one object must equal the total heat entering another. Third, heat flow is something we only consider if

---

* As an example, for monatomic gasses, like Helium, their internal energy is equal to $\frac{3}{2}NkT$, where N is the number of gas particles, k is the Boltzmann constant, and T is the absolute temperature in Kelvin.

the two objects are at different temperatures. You can have liquid water at 0 °C in contact with ice at 0 °C, and no heat would flow between them.

The First Law of Thermodynamics

As in any other system, the total energy is conserved. This is applied in the First Law of Thermodynamics, where we look at how energy transfers from one object to another. The First Law of Thermodynamics is that the change in internal energy, ΔU, of a system, is equal to the total heat flow, Q, INTO the system plus the total work, W, done TO the system (W).

$$\Delta U = Q + W$$

This equation is deceptively simple, as we must be careful about the *signs* in it. On the right-hand side, positive values correspond to energy entering the system, and negative values correspond to energy leaving the system. Positive heat flows mean heat is entering the system, and negative heat flows mean heat is flowing out of the system. For Work, positive values correspond to something external acting on the system, applying forces over distances. Negative values for Work correspond to the system is using energy to do Work, so Work done BY the system. Compressing a gas is considered positive Work, as the net force is against the gas, and a gas expanding is considered to be negative Work, as the gas presses on the walls of its container.

A common case study that used to examine the First Law of Thermodynamics is gas sealed within a beaker with a moveable piston top. As long as the beaker is not insulated, heat can flow into or out of it. If the top of the beaker is moveable, that means work can be done to or by the gas.

Let's say that we light a fire under this beaker. This causes heat to flow into the gas. At the same time, let's say that the piston top of the beaker moves upward, meaning that the gas inside expands. The gas pushes the piston top upward, thus exerting a force over distance, and work is done by the gas. If 50 J of Heat energy flows into the gas, and the gas does 30 J of Work moving the piston up, then the change in Internal Energy of the gas is +20 J.

$$\Delta U = Q + W = (+50\,J) + (-30\,J) = +20\,J$$

Transferring Heat

Let us return to Heat and examine how it flows from one object to another. Heat flow is usually categorized into Conduction, Convection, and Radiation. Realistically, all three forms of transfer are active, but breaking heat flow down in this way allows us to describe the time behavior of this energy transfer.

Conduction is the transfer of Heat through solid objects. A practical example of this is in a window or the walls in your home. On most days, the temperature inside your home is quite different from the temperature outside, so heat is going to flow through the walls and windows until they are balanced. However, this process takes more or less time depending on what materials are used. This conduction of heat can be defined explicitly through the following equation:

$$\frac{Q}{t} = \frac{kA}{d} \Delta T$$

where Q is the heat flow, t is the time, ΔT is the change in temperature, k is the thermal conductivity of the material, A is the cross-sectional area of the material, and d is the material's thickness. In construction, the thickness and thermal conductivity are grouped together to define the R-value of that material[*].

Convection is the transfer of heat through the motion of particles in fluid. It is a combination of advection (bulk-flow) and diffusion. Unfortunately, the fundamental math to describe these processes exactly are trickier, as they are partial differential equations involving both spatial and time terms[†]. A good example of Convection is a fireplace heating

---

[*] To be precise, $R_{Value} = d/k$

[†] $\frac{\partial T}{\partial t} = \nabla \cdot (D \nabla T) - \nabla \cdot (T\vec{v})$ where D is the Diffusion Coefficient for the substance and $\vec{v}$ is the velocity. In

a room. Over time, different packets of air get in contact with the fire, causing their temperatures to increase. Those now hot packets of airs move to other places in the room, increasing the temperatures of the air around them, while other packets of air also get warmed by the fire. Over time, all of the air in the room gets warmed up by the fire.

The third category of heat flow is **Radiation**, the transfer of heat via light or electromagnetic radiation. All objects do this – it's just that amount and frequencies of light depend on the objects' temperatures. Things have to be several hundred degrees for their light to be emitted in the visible range. Infrared cameras function on this principle. People and other living creatures are usually warmer than their surroundings. Thus, they emit more infrared light than the objects around them, causing them to show up on infrared cameras.

This process is also known as Blackbody Radiation. There is a characteristic curve describing the amount and frequencies of light based on an object's temperature, as shown below.

---

layman's terms, the change in temperature over time is equal to how the particle temperatures and velocities vary with position. If you want the overall effect for the object as a whole, you can use Newton's Law of cooling, $Q = Ah\Delta T$, where A is the area of the object and h is the heat transfer coefficient for the system

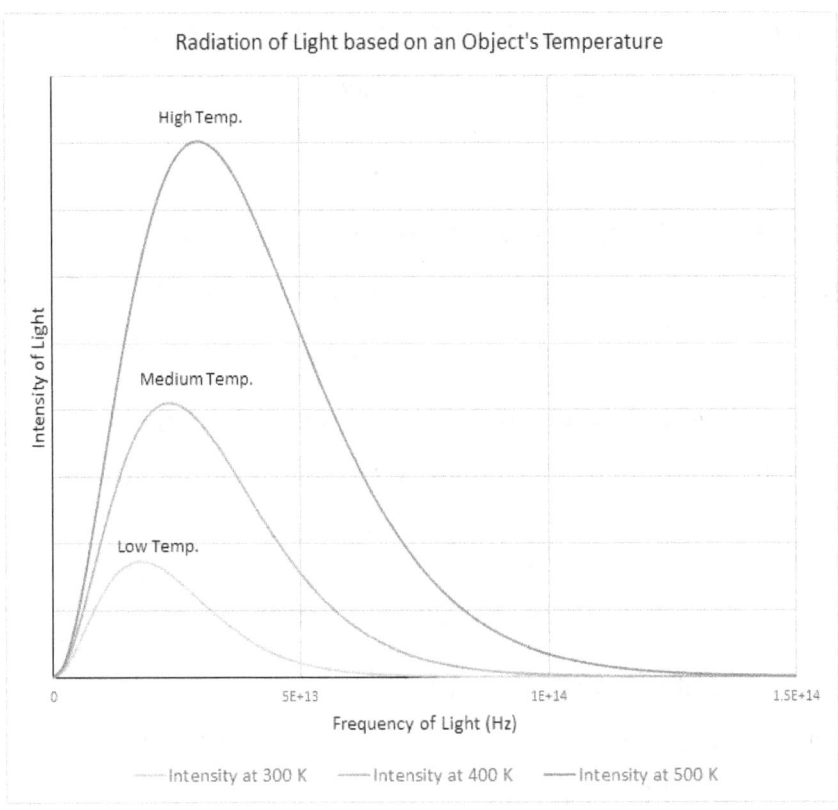

These were experimentally determined in the mid-to-late 19th century. It turns out that the correct theoretical derivation for these curves was a starting point for Quantum Mechanics and will be revisited in a later chapter. The equation to describe the intensity of light at a given frequency and temperature is a bit messy*, but the total heat

---

* The Spectral Radiance: $B(f,T) = \dfrac{2hf^3}{c^2} \cdot \dfrac{1}{e^{\frac{hf}{k_B T}} - 1}$ where B is the Intensity per solid angle and area, h is Planck's constant, c is the speed of light, f is the frequency, k_B is the Boltzmann constant, and T is the temperature in Kelvin.

flow over time is relatively straightforward. This is also known as the Stefan-Boltzmann law.

$$\frac{Q}{t} = \sigma A T^4$$

where Q is the heat flow, t is the time, A is the area of the object, T is the temperature in Kelvin, and σ is the Stefan-Boltzmann constant and is equal to $5.67 \cdot 10^{-8} \frac{W}{m^2 K^4}$.

If you want to find out the peak wavelength of light emitted by an object at a given temperature, you use Wien's Law:

$$Peak\ Wavelength\ (m) = \frac{2.898 \cdot 10^{-3}\ m \cdot K}{Temperature\ (K)}$$

For example, in order for the peak wavelength to hit the visible range (~740 nm), the temperature of the object needs to be about 3900 K!

Results of Heat Transfer

As heat is transferred into or out of a system, two things can happen: a change in temperature and/or a change in phase. The equation describing the heat causing a temperature change is as follows:

$$Q = mC\Delta T$$

where Q is the heat flow, m is the mass of the substance, ΔT is the change in Temperature, and C is the Specific Heat Capacity of the substance. The **Specific Heat Capacity** is a critically important variable, as it basically tells you how a substance is going to react Thermodynamically. It is a property of a material and its phase that tells you how much energy is needed to change its temperature. Things like metals have a very low specific heat capacity (around $500 \frac{J}{kg \cdot K}$), meaning that it takes very little energy to cause its temperature to vary. Liquid water, however, has a very high specific heat capacity (around $4180 \frac{J}{kg \cdot K}$), meaning that it takes a tremendous amount of energy to change water's temperature. This is a major determinant in the range of temperatures in a particular region. Deserts, which have very little water around, get to be very hot during the day and very cold at night. However, coastal regions and regions with large bodies of water around them do not experience such extremes, since all of that water helps to regulate the temperature.

If additional heat is transferred into or out of a system, that system may also experience a change in phase. The equation governing this process is as follows:

$$Q = mL$$

where Q is the heat flow, m is the mass of the substance, and L is the latent heat. The **latent heat** is a property of the material and whatever phase transition is occurring: fusion for solid ↔ liquid, vaporization for liquid ↔ gas, or sublimation for solid ↔ gas. Changing an object's phase is basically a reformation of the atomic bonds present, and requires a lot of energy to do so. For example, the latent heat of fusion for water is about 334,000 J/kg, and the latent heat of vaporization for water is about 2,264,000 J/kg. To give you a sense of scale, the energy needed to raise 1 kg of water's temperature from the freezing point of 0 °C up to the boiling point of 100 °C is about 418,000 J. The energy needed to vaporize that now 100 °C water is roughly five times of that. This illustrates why water is so useful in cooling things.

Entropy and the Second Law of Thermodynamics

The last topic to discuss in Thermodynamics is **Entropy**. Unfortunately, the way it is commonly described or used in everyday language makes it seems like some abstract hidden force of destruction. In reality, Entropy is nothing more than reflection of the statistical nature of Thermodynamics. The standard definition of Entropy is that it is the relative disorder of a system, but that's not particularly helpful, as "disorder" is a vague, subjective term. Mathematically, Entropy is related to the number of ways a system's particles can be arranged*. To give you an idea of how it works, let us go through the following analogy.

Imagine that we have a 2x2 grid and a white marble. We can place that white marble in any of the four slots available. This is akin to a low temperature and low entropy system. Apart from that, we have a second 4x4 grid and a black marble. We can place that black marble anywhere on that second grid, meaning that it has 16 slots available. This is akin to a high temperature and high entropy system. If we consider the total number of options available to both marbles in their respective grids, we have to multiply their respective options (4 and 16), giving us a total of 64 possibilities for those two marbles.

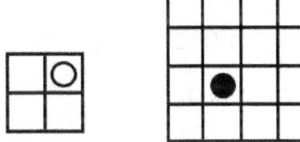

Now, let us say we combine those two systems into a single 3x3 grid. This is like letting heat flow between them. Starting with the black marble, the black marble has 9 slots available. Since the black marble takes one slot away, the white marble has only 8 slots available. Taken

---

*To be precise, Entropy (S) is the Boltzmann Constant (kB) times the natural logarithm of the number of microstates of a system (Ω). That famous equation is on Ludwig Boltzmann's tombstone: $S = k_B \ln \Omega$

independently, the black marble has lost 7 options, and the white marble has gained 4. However, to get the *total* number of options, we multiply their 9 and 8 options together, giving use 72 possibilities for both of the marbles in combination – an overall net increase from the 64 possibilities from before.

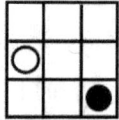

This illustration is basically the Second Law of Thermodynamics in action – the combined Entropy of two interacting systems can only increase or remain constant. When two systems interact, one part may lose options, and another part may gain options, but the net effect is an increase in the number of options available for those systems' particles. The only way the combined entropy will remain constant is if the two systems stay relatively isolated and no heat flow is allowed.

If a high energy system becomes in contact with a low energy system, energy will be exchanged between them until everything ends up at around the same temperature. The reason why this happens is that there are simply more options where the energy is evenly distributed versus the number of options where the energy is concentrated into a single spot.

For example, the particles of air are distributed throughout the room. There is a possibility that all of the air becomes concentrated in one corner and you begin to suffocate. However, that is not something you need to worry about since there are far, far more options for the air particles to be distributed evenly versus concentrated in one particular part of the room.

Notice that there was no "physics" or additional forces included in the above discussion. Again, Entropy just reflects the statistics of how things interact. A good rule of thumb is that any time there is heat flow (an exchange of energy), the two systems interact, energy gets distributed, and the combined entropy of the two interacting systems increases[*].

Entropy can be a bit tricky understand. An excellent resource for the math of Entropy and how it works is "Entropy Demystified: The Second Law Reduced to Plain Common Sense" by Arieh Ben-Naim. It is written for a general audience as well.

---

[*] Mathematically, the incremental heat flow is equal to the absolute temperature times the incremental change in entropy: $\delta Q = T \delta S$

Ch. 5 – Fluid Principles

Defining Fluids

Liquids and gasses are both considered to be fluids, as their atoms can freely move within their volumes. In this chapter, we will look at some additional properties and applications that are unique to them. This will be a macroscopic approach where we treat the fluids as continuous mediums. Additionally, the fluids will be considered ideal, meaning that they are incompressible (have constant density), non-viscous (fluid's version of friction), and flowing at a rate less than the speed of sound.

Law of Fluid Pressure

The first fluid principle we will observe is a result of gravity. Imagine that you are a diver in a pool or the ocean. As soon as you have gone down a certain distance, some amount of water is above you. Gravity is acting on that water, pushing it down. As the water pushes down on you and its lower layers, that push from gravity gets translated into an added pressure. Thus, the deeper you go in a fluid, the greater the pressure. This can be applied to air in the atmosphere as well. The higher you go, the lower the air pressure, since there is now less amount of air being pressed down. This pressure difference is then added to the atmospheric pressure at sea level to get the total pressure.

Mathematically, the law of fluid pressure states that this added or gauge pressure, P, is equal to the density of the fluid, ρ, times the acceleration due to gravity, g (which is still 9.8 m/s²), times the relative depth or height change, h, in the fluid.

$$Added\ Pressure = \rho g h$$

As an example, let's look at how the pressure changes as a diver goes down 100 meters in the ocean. The density of salt water is about 1202 kg/m³. This means that the added pressure at that depth is:

$$Added\ P = \rho g h = \left(1202 \frac{kg}{m^3}\right) \cdot \left(9.8 \frac{m}{s^2}\right) \cdot (100\ m)$$
$$= 1{,}177{,}960\ Pa$$

This is over ten times the atmospheric pressure of 101,000 Pa! Because water is so dense, the pressure adds up quickly the deeper you go, which is why divers need to be careful when ascending or descending.

Archimedes' Principle

The next principle is also a result of gravity. Let us continue to picture that diver in the water. As that diver moves into the water and below its surface, it pushes away or displaces a certain amount of water. That amount of water is being pulled down by the force of gravity, but its original position is occupied by that diver. By going into the water, the diver is pushing against gravity acting on that amount of water. By Newton's 3rd Law of Motion, if the diver pushes the water away, the water pushes back with an equal and opposite force.

This push back from the water is called the buoyant force and is described by Archimedes' Principle. The buoyant force is equal to the force of gravity on the displaced fluid. The force of gravity is the mass times the gravitational acceleration, g. The mass of the fluid in this case is the density of the fluid times the volume that was displaced. In equation terms, the buoyant force, F, acting on an object from a fluid is equal to the density of the fluid, ρ, times the displaced volume, V, times the acceleration due to gravity, g.

$$Buoyant\ Force = \rho_{fluid} V_{displaced} g$$

In the case of a diver being pushed by water or a person on land being pushed by the air in the atmosphere, the volume being displaced is that person's whole volume. However, for objects that are floating on the water, only *some* of that object's volume is needed for the buoyant force to fully counteract the force of gravity on that object. This principle can be used to find how of the ship's volume is below the water line*.

Pascal's Principle

This next principle is very much a special case, as it only applies to ideal fluids that are at rest and completely fill an enclosed volume. However, it has an important application in the form of hydraulic lifts. Pascal's Principle states that for a fully enclosed fluid at rest, any added pressure at one location in the fluid is transmitted equally and undiminished throughout the fluid. In other words, pressure that is added in one spot gets added in all spots.

Let us see how this gets applied in a simple hydraulic lift. For this basic setup, a container is filled with water and has two moveable piston tops. One side, the input, has a small area piston, and the other side, the output, has a larger area piston. This is illustrated in the following figure.

---

*In terms of equations, the buoyant force must cancel out the force of gravity on the object, so $F_{Buoyant} = F_{gravity}$ → $\rho_{fluid} g V_{displaced} = m_{object} g$, therefore
$$V_{displaced} = \frac{m_{object}}{\rho_{fluid}}$$

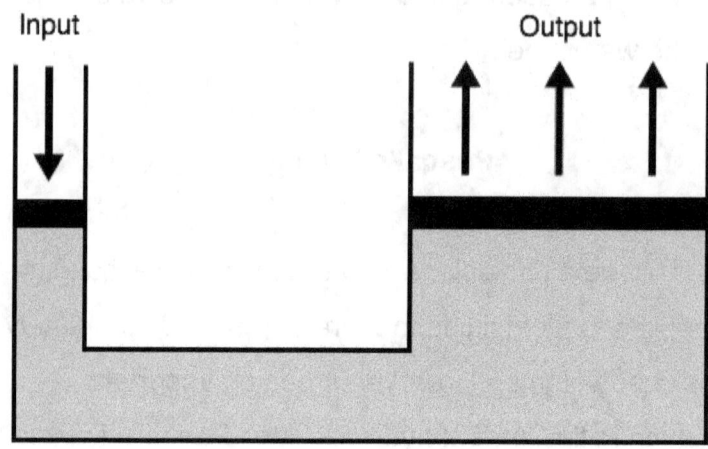

Next, a small force is applied to the input side, which adds a certain amount of pressure at that location. Pascal's Principle states that this added pressure is transmitted throughout the fluid. This means that the same amount of pressure is added to the output side of the system.

$$\text{Added } P_{In} = \text{Added } P_{Out}$$

Because pressure is defined as the force applied over an area, this means that the input force divided by the input piston area must equal the output force divided by the output piston's area.

$$\frac{\text{Added } F_{In}}{A_{In}} = \frac{\text{Added } F_{Out}}{A_{Out}}$$

Therefore, a small force applied over a small area translates into a large force over a large area! This an example on how shifting a certain amount of water can end up lifting heavy objects. That being said, there are some limitations to Pascal's Principle. In addition to fluids not necessarily being ideal, energy conservation must be applied. By pressing down on the input side over a distance, work is being done. Due to the conservation of energy, the work that is done in moving the fluid down on the input side must equal the work being done to lift the fluid on the output side.

$$Input\ Work = Output\ Work$$

Work is defined as the force applied times the distance. In this case, the small input force needs to be applied for a large distance in order for the larger output force to be applied for even a short distance. This ends up being the same relation that holds for the simple machines described in a previous chapter.

$$Small\ Input\ Force \cdot Large\ Distance$$
$$= Large\ Output\ Force \cdot Small\ Distance$$

Continuity Equation

Aside from nuclear reactions, mass is a conserved quantity. When applied to fluids, the total amount of fluid entering a region must equal the total amount of fluid exiting that region – what comes in must goes out. This is called a Continuity Equation*.

For a non-leaking fluid, the mass flow rate is constant, where the mass flow rate is defined as the density of the fluid, ρ, times its speed, v, times the area it is moving through, A. This mean that the fluid's speed and the area it moves through are inversely proportional.

$$Mass\ Flow\ Rate = \rho v A = constant$$

You can see this in action with water coming out of a hose. If there is no cover on the end of the hose, water fills the cylindrical area and comes out at a slow speed. However, if you place your thumb over part of the end of the hose but still allow the water to flow, water jets out quickly through the small gaps that are left. By reducing the area the water can move through, its speed must increase to compensate. This is because the same amount of matter is coming out of the hose no matter what!†

Bernoulli's Principle

In the last chapter, we stressed the importance of the conservation of energy. When applied to fluids, it is called

---

* A similar continuity equation exists for charges and current flow.
† As long as the fluid is allowed to flow and not completely blocked, that is.

Bernoulli's Principle. Its results can be counter-intuitive, so it is worth doing a quick derivation from a conservation of energy equation. To make things simpler, we'll continue to treat the fluid as ideal, so we don't need to worry about the energy associated with viscosity quite yet. For a fluid flowing through a certain space, it has a certain amount of kinetic energy. As long as it's on Earth, there is also some gravitational potential energy. Lastly, we know that the fluid particles interact with each other via collisions at a minimum, so let us include a generic potential energy or work term. Thus, the total energy of the fluid is the sum of those three forms, and that sum must be constant.

$$Total\ Energy\ of\ the\ Fluid =$$
$$= Kinetic\ Energy + Gravitational\ Potential\ Energy$$
$$+ Work$$
$$= constant$$

Next, let us insert our previous equations for Kinetic Energy (1/2 times the mass times the speed squared), Potential Energy (mass times the acceleration due to gravity times height), and Work (Force times distance).

$$Total\ E = \frac{1}{2}mv^2 + mgh + Fd = constant$$

When dealing with fluids, we most often deal with them in terms of volume, not necessarily in terms of their total mass. So, let us divide each term in the above equation by the volume of the fluid, V.

$$\frac{Total\ E}{V} = \frac{\frac{1}{2}mv^2}{V} + \frac{mgh}{V} + \frac{Fd}{V} = \frac{constant}{V}$$

The sum of these quantities is still a constant – it just has a different value now. For the kinetic energy and gravitational potential energy terms, the mass is divided by the volume, and we can replace those two variables with the density of the fluid. What about the term involving work? We didn't specify anything regarding the *shape* of this volume. We can safely rewrite the volume as some area times a length or distance.

$$\frac{Fd}{V} = \frac{Fd}{Ad}$$

Note that the distances cancel out on the top and bottom, and we are left with the force of the interatomic interactions divided by the area of the fluid. This is the same thing as its internal pressure, P! With that in mind, let us rewrite our adjusted energy equation with those replacements. This is Bernoulli's Principle.

$$\frac{1}{2}\rho v^2 + \rho gh + P = constant$$

For fluids, the internal pressure functions like a potential energy term. If a fluid is kept at a constant elevation (meaning h is constant), then the fluid's speed and pressure are inversely proportional to each other. For example, if the speed of the fluid increases, then its internal pressure decreases. As the fluid's kinetic energy goes up, its potential energy must go down, which it does in the form of a lower pressure.

Part of the problem when discussing these results is that we often use "pressure" in everyday language to describe a fluid's speed. The speed can be described in terms of kinetic pressure (how hard it is hitting stuff). The pressure described in all of the previous equations is the Static Pressure – how the fluid particles interact with each other, and this can be inverse of the speed. What happens is that if the fluid's atoms are moving much faster in one direction, they are not colliding and interacting as much in the other directions. This results in a lower internal static pressure.

An important application of Bernoulli's Principle is the Lift for airplanes. Airplane wings are designed to have air flow faster on top of the wings and slower below. Because there is a difference in speeds between the top and bottom, there is also a difference in air pressure. On top, where the air's speed is faster, the wing experiences a lesser pressure pushing it down, as compared to the bottom of the wing where the air speed is slower and the upward pressure is greater. Through this difference in pressure, the wing experiences a net upward force, giving the airplane Lift!

Viscosity

Many fluids, like oil or molasses, have significant viscosity. This greatly influences their motion in a number of ways. One effect is turbulence, where the fluid's speed greatly varies in both position and in time. Solving for turbulence is extremely difficult and is a longstanding problem in Physics. A second effect of viscosity is called laminar flow. Basically, the viscous fluid travels in parallel layers. If the friction or viscosity between the container and fluid is greater than the viscosity within the fluid, then the layers of the fluid in contact with the container move slower than the inner layers of the fluid. The reverse can happen as well. In either case, the energy transfer due to viscosity results in a decrease of the fluid's kinetic energy, where different parts of the fluid are affected at different times. If, however, the fluid completely fills the volume and maintains a constant speed, there is still some energy transfer due to viscosity. In this case, it results in a decrease in the fluid's pressure.

Ideal Gas Law

The last fluid principle we will discuss applies only to gasses, and has very important applications in engineering, physics, and chemistry. It is called the Ideal Gas Law. Ideal gasses are those whose particles only interact via collisions (meaning there are no chemical interactions), and most gasses and gas mixtures, including air, can be considered ideal. Ultimately, this is another application of the conservation of energy. What happens is that the number of gas molecules, N, the gas's absolute temperature, T (which must be in Kelvin), total pressure, P, and volume, V, are all connected via the Boltzmann Constant, k. In standard metric units, the Boltzmann Constant is equal to $1.38 \cdot 10^{-23}$ Joules per Kelvin (J/K).

$$PV = NkT$$

Most often, this law is applied for a fixed number of gas particles. In this case, you can relate the initial values of the total pressure, volume, and absolute temperature to their final values quite easily.

$$\frac{P_0 V_0}{T_0} = \frac{P_F V_F}{T_F}$$

An important example of how this is used is in combustion engines. During the combustion, the temperature of air rapidly increases. This increase in its temperature results in an immediate increase in pressure, which then cause the pistons in the engine to move. Thus, chemical energy is transferred to mechanical energy.

Ch. 6 – Waves

Defining Waves and Their Components

Waves are a special class of motion and interactions that exist in systems. They form a basis of many kinds of physical phenomena and play a major part in modern physics. In this chapter, we will look at what makes up a wave, a few of their major properties, and examine sound waves as a useful case study.

Waves can be defined as disturbances that travel through a medium. An important thing to keep in mind is that waves oscillate both in terms of time and position. If you want to know what a wave is doing, it depends on both where you look and when you look. Additionally, this means that you need at least two graphs if you want a full picture of the wave – one graph for the disturbance in terms of time at a given location (say, at x = 0, x = 5 m, and so on), and one graph for the disturbance in terms of position at a given time (t = 0, t = 10 s, etc.). The following two graphs show a sample wave that will be referenced in this section. We'll come back to their equations at the end.

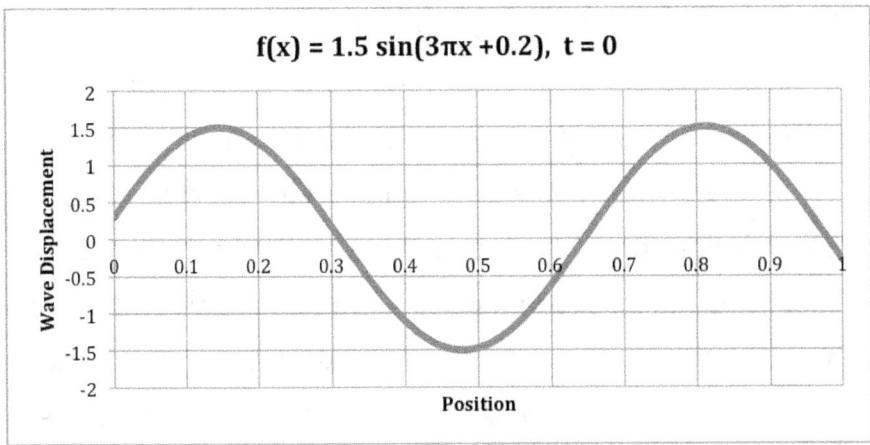

Plotting the wave's displacement in terms of position at t = 0.

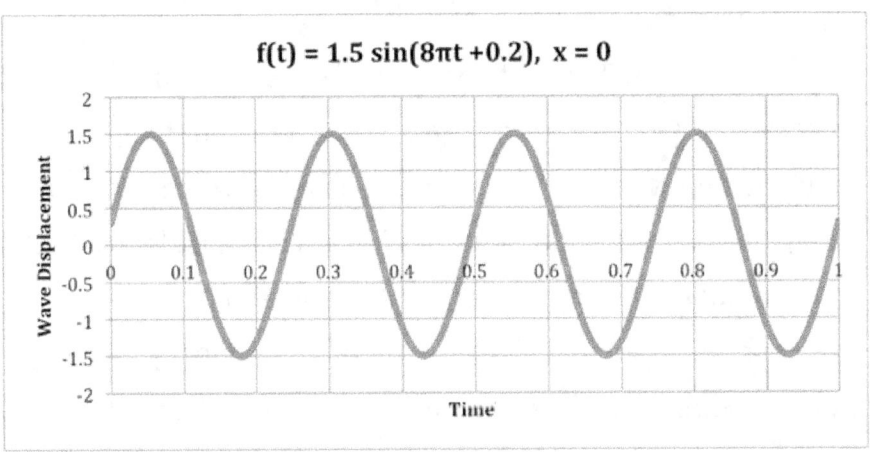

Plotting the wave's displacement in terms of time at x = 0.

The relative strength of that oscillation or maximum disturbance (either positive or negative) is called the **Amplitude**, A. In the above graphs, the amplitude is 1.5. Again, waves can oscillate in both time and space. In more complex waves, the Amplitude can be a vector. Locations and times where the displacement of the particles is zero are known as *nodes,* and locations and times where the displacement of the particles is equal to the amplitude of the wave (either positive or negative), are known as *antinodes.*

The distance it takes to go through one cycle is called the **wavelength**, λ. On the first graph, this would be the spatial distance between two positive peaks or between two negative peaks. In our graph for the wave's displacement vs position, the wavelength is 2/3 m. For waves in two or three dimensions, there are potentially different wavelengths for each direction.

The time it takes to go through one cycle is the **period**, T. On the second graph, this would be "temporal distance" between two positive peaks or between two negative peaks. The **frequency**, f, of a wave is defined as the number of cycles divided by the time and is the inverse of the period (so f = 1 / T). The units for frequency are cycles per second (1/s), which are also called Hertz, Hz. In our graph for the wave's displacement vs time, the period is 1/4 s and the frequency is 4 Hz.

The starting point or time of the wave does not necessarily have to correspond to the wave's displacement equaling zero or the maximum displacement. Those different "starting points" are mathematically described by the **phase angle**, ϕ, of the wave. In our graphs, the phase angle is equal to 0.2.

When we construct a mathematical equation to define what the wave is doing at a particular time or location, we use the trigonometric sines and cosines*. In general, a one-dimensional wave function can be formulated as follows†:

$$f(x,t) = A \sin\left(\frac{2\pi x}{\lambda} - \frac{2\pi t}{T} + \phi\right)$$

The wave used in the above graphs would be written as:

$$f(x,t) = 1.5 \sin(3\pi x + 8\pi t + 0.2)$$

Most waves we encounter are a combination of several individual waves, each with their own amplitudes, phase angles, frequencies, and wavelengths. Additionally, *all* waves can be broken down to a sum of those individual waves‡.

---

* For those who unfamiliar with trigonometry, sines and cosines can be thought of as ratios for the sides of triangle. If you've had some calculus, you can also define the waves as complex exponentials $(e^{i\theta})$

† The partial differential wave equation that creates this function is $\frac{T^2}{\lambda^2}\frac{\partial^2 f}{\partial t^2} = \nabla^2 f$

‡ This is because sine and cosine functions form a complete mathematical set. You can define *any* function as a sum of sine and cosine functions with different arguments.

How Waves Travel and Interact

The speed at which a wave travels can be found simply by combining its position behavior to its time behavior.

$$Wave\ Speed = Frequency \cdot Wavelength$$
$$v = f\lambda$$

Like any other speed, it has units of m/s.

The specific term that we use to describe waves travelling is *propagation*. If the wave is travelling or propagating in the same direction as its oscillation, it is called a **Longitudinal** Wave. If the wave is propagating in a direction perpendicular to its oscillation, it is called a **Transverse** Wave. Most waves, such as those we see at the beach, are a combination of longitudinal and transverse waves. As a wave travels through a medium, it's maximum oscillation or amplitude can decrease, which is called *damping*. The medium can also absorb some of the energy of the wave.

Once a wave hits a new material or some kind of boundary, it can either be reflected, like light hitting a mirror, or transmitted, like light passing through a window. There are some special cases associated with both of these. If the wave reflects back and forth in a set region, a standing wave is formed. Reverberation is an example of this. As a wave transmits into a new material, the path of a wave can change. This is called *refraction* and will be revisited in a later chapter for light waves in particular.

When two or more waves are combined in some fashion, it is called **interference**. That interference is *constructive* if the waves align and add together, and *destructive* if the waves cancel each other out to some degree. Both kinds of interference occur at set intervals when waves pass through one or more gaps. This is called *diffraction* and will also be revisited later for light waves.

If two waves that have a very similar frequencies combine, an important pattern emerges. A rapid oscillation occurs within a general envelope with a **beat frequency** equal to the difference of the two individual frequencies. The figure below illustrates the combination of two waves, one with a frequency of 20 Hz and another with a frequency of 22 Hz. The envelope or beat that is formed has a frequency of 2 Hz (22 – 20). When one is tuning an instrument, the pulsing sound one hears is that beat or envelope frequency, which is the difference between the frequency of the instrument and the frequency it is tuning against.

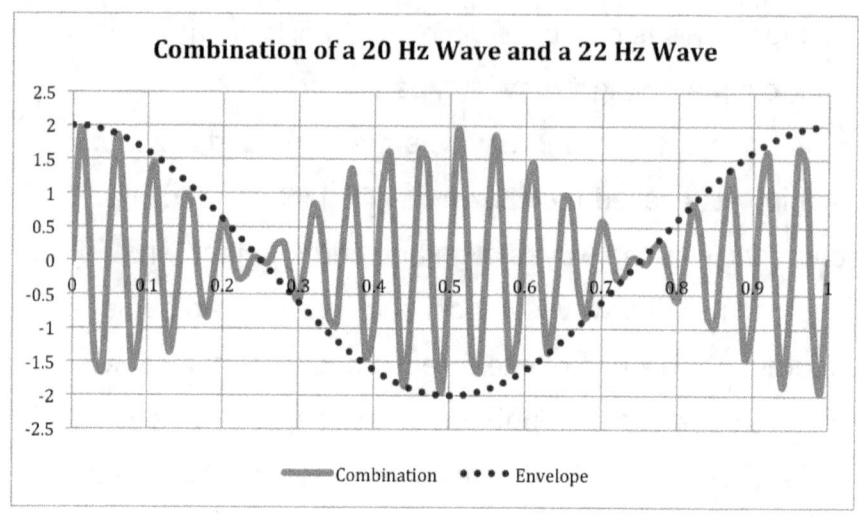

Sound Waves

Sound waves are variations in pressure through a medium and are a useful case study for observing many wave principles. It propagates longitudinally through compressible mediums (such as air, water and other kinds of fluids), but both longitudinally and transversely through solids. The speed of sound depends on the type of material, the phase of matter, the temperature, and other factors*. For example, the speed of sound through air at room temperature is about 343 m/s, and the speed through body tissue is about 1540 m/s.

---

* Mathematically, it is the square root of the partial derivative of the pressure with respect to the density: $v = \sqrt{\left(\frac{\partial P}{\partial \rho}\right)}$

When we evaluate the amplitude of a sound wave at a given point, we consider its **Intensity**, I, which is defined as the power of the wave exerted over some area and has units of W/m². This intensity is based on the pressure difference occurring, the density of the material, and the speed of sound in that material,

$$Intensity = \frac{Power}{Area} = \frac{(Change\ in\ Pressure)^2}{2 \cdot Density \cdot Speed}$$

Humans can hear frequencies of sound from about 20 Hz to 20 kHz. Most sounds that we hear are a combination of multiple frequencies. An important factor in how we distinguish different sounds are the harmonics that are present or absent, where **harmonics** are integer multiples of some base frequency. For example, the harmonics of a 500 Hz tone are 1000 Hz, 1500 Hz, 2000 Hz, and so on.

The perceived "loudness" of sound is related to both the intensity and frequency of the sound wave. Our ears are more sensitive in the 2000 – 5000 Hz range. Greater intensities of sound are also interpreted to be louder. Furthermore, this "loudness" from the intensity is interpreted *logarithmically*, meaning that it based on powers of ten. The decibel scale is how we evaluate this logarithmic relationship. Here, the relative loudness in decibels is based on the logarithm of the intensity of the sound wave over the baseline intensity, where the baseline intensity, I_0, is the smallest intensity of sound that humans can reliably hear at about 10^{-12} W/m².

$$Relative\ Loudness\ in\ Decibels = 10\ \text{dB} \cdot \log_{10}\left(\frac{I}{I_0}\right)$$

Each power of 10 in intensity is thus 10 dB and is perceived to be 2 times louder. For example, if one sound wave has 1000 times the intensity of another, it is 30 dB more (10^3 = 1000), and is perceived to be 8 times louder (2^3 = 8).

One important wave property that we normally associate with sound is the Doppler Effect. The Doppler Effect is the shift in the frequency of a wave due to the motion of a wave's source or receiver. When objects move closer to each other, the wave fronts become closer together, resulting in a higher frequency. Likewise, when objects move further away from each other, the wave fronts become more separated, resulting in a lower frequency. You can hear this as a siren approaches you and then passes you by.

The Doppler Effect for light is how we found out that the universe is expanding. We know what stars are made of, and know what frequencies of light should be emitted from them. However, in nearly all stars, those frequencies are measured to be lower than expected. This means that those stars are moving away from us. Since this is occurring in all directions, the space between things must be increasing, and thus the universe is expanding.

The sonic boom is the sound associated with the shock waves created by an object travelling at least as fast as the speed of sound in air. It is basically an extension of the Doppler Effect. When the object is moving at least as fast as the sound waves, those wave fronts begin to stack on each other. Since sound waves are variations of pressure, this stacking of wave fronts results in an enormous change in pressure and high intensity (about 110 dB). The sonic boom that we hear is that sudden gain and loss of a high amount of pressure. When an object is moving faster than the speed of sound, these high/low pressure regions form a "V" shape, with the apex at the tip of the object.

Many of these wave principles are difficult to visualize in text or static pictures. An excellent resource for animations and accompanying explanations can be found at https://www.acs.psu.edu/drussell/demos.html. These were written and developed by Dr. Daniel A. Russell at Pennsylvania State University.

Ch. 7 – Electricity and Magnetism

Charge and Current

So far, every quantity discussed has been some combination of the base measurements of mass (kg), length (m), and time (s). There is another fundamental variable and unit to consider, and that is **Charge**, using the symbol of q or Q, and having the units of Coulombs, C. Charge is an intrinsic property of certain subatomic particles. Just as a particle may have a certain amount of mass, it may have a certain amount of charge. Unlike mass, the charge of a particle can either be positive, negative, or zero. Furthermore, charges exist in discrete amounts. For right now, the smallest amount of charge we'll consider is that of a proton or an electron. Both have the same magnitude of charge, e, equaling $1.6 \cdot 10^{-19}$ C, where the proton has a charge of +e and an electron has a charge of -e. Protons are roughly 1800 times more massive than an electron, and are primarily fixed in the nuclei of the atoms of a substance. Electrons, however, are more-or-less free to move around between those atoms. Barring any outside influence, the atoms in a material have an equal number of protons and electrons, and therefore have no net charge.

The rate at which charges move is called **current**, I. It has units of Amps, A, which are the same as C/s. However, current is usually measured in mA, since most circuits operate on that scale.

$$Current = \frac{Change\ in\ Charge}{Change\ in\ time}$$

$$I = \frac{\Delta q}{\Delta t}$$

The Electric Force

Charges are involved in both electric and magnetic effects. The primary interaction between charges is the electric force. Like any other force, the electric force is a vector having both magnitude and direction. For the direction, like charges repel, and unlike charges attract. This means that two positive charges want to get away from each other, two negative charges want to get away from each other, and a positive charge and a negative charge are attracted towards each other.

Coulomb's Law gives the magnitude of the force between two charges. It is based on the value of each charge (q_1 and q_2), the distance between them (r), and a scaling constant, k. The Electric Constant, k, has a value of about $9 \cdot 10^9 \ N \cdot \frac{m^2}{C^2}$.

$$|F| = \left|\frac{k q_1 q_2}{r^2}\right|$$

The electric force is an example of an inverse square law, where the force is proportional to one over the distance squared. Inverse square laws are commonplace in physics. The closer the charges are to each other, the stronger the force, and the strength of the force decreases rapidly with distance.

An important feature of the electric force is its strength. When charged objects are present, the electric force generally will dominate over everything else. This is especially true at the particle level. For example, the electrical force between a proton and an electron is about 10^{39} times greater than the gravitational force between them!

For an object that has zero net charge, there can be no net electrical force acting on it. However, charges *outside* the object can still cause a noticeable force on the electrons of that neutrally charged object, causing the electrons to shift slightly to one side, thus *polarizing* that object. This effect is akin to the gravitational effect of the Moon pulling on Earth's oceans, causing tides to develop. The following figure illustrates this principle.

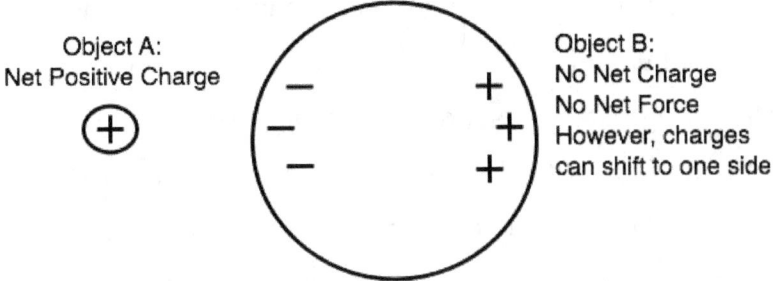

The Electric Field and Potential

Normally, when dealing with electrical interactions, we have a collection or distribution of charges, not individual point charges. Alternatively, we have one set of charges (set A) based on the equipment being used that can interact with some other charges that are introduced and swapped out (sets B, C, D, and so on). In these scenarios, it is helpful to introduce a new tool to describe how a collection of charges influence their environment. We'll do so with the **Electric Field**, E, the vector that describes all of the electric effects due to the charges that are present in an environment. The Electric Field has units of N/C. We'll use the Electric Field to describe the Electric Force that a charge experiences once it is introduced to that environment. The Electric Force, F, on a charge, q, is equal to the value of that charge times the Electric Field, E, at its location.

$$F = qE$$

This may seem to be a bit of arcane mathematical trickery, but we have already encountered something like that for gravity: *g*. In essence, *g*, the free-fall acceleration near the surface of Earth, describes the *gravitational* effect of Earth on the environment around it. The gravitational force on some new added mass is equal to its mass, *m*, multiplied by the gravitational field at that mass's location: *F = mg*! This same principle can be applied to electricity.

The way we construct the electric field is essentially by taking Coulombs Law ($|F| = \left|\frac{kq_1q_2}{r^2}\right|$) and removing one of the charges. For the direction, the Electric Field points away from positive charges and towards negative charges. The magnitude of the electric field from a point charge is equal to the value of the charge, q, times the electric constant, k, divided by the square of the distance from that charge, r.

Electric Field from a Point Charge: $|E| = \left|\frac{kq}{r^2}\right|$

To find the total electric field at a given location, you simply add up the electric field contributions from every charge present in that environment. For more complex arrangements of charges, you have to employ some calculus to get the overall effect[*].

At the start of the text, we described motion using forces and vectors, then we moved on to describing motion

---

[*] It still boils down to adding those individual fields, albeit in differential form.

with energy and scalars. For the primary electrical interaction, we can also describe its effect in terms of energy and scalars. First, we do a scalar treatment of the electric field. The **Electric Potential**, V, is the <u>scalar</u> that describes all of the electric effects due to the charges that are present in a given environment. It has units of Volts, which also use V as its symbol, so Voltage and Electric Potential are used interchangeably. For example, in a 9 V battery, that 9 V is telling you the electrical effect due to the positive and negative ions being separated in that battery.

Just as we used the Electric Field to describe the force on a charge, we can use the Electric Potential to describe the energy of a charge. The electric potential energy of a charge is equal to the value of the charge times the electric potential at its location. Unfortunately, we run into a bit of a language issue and symbol issue – the symbol for the electric potential is V and the symbol for the potential energy can also be V. The *Electric Potential Energy* is how much electrical energy a particular charge possesses. The *Electric Potential* or *Voltage* is what is going in the background due to everything else.

The way we construct the electric potential is similar to how we construct the electric field. Fortunately, the electric potential is a scalar, so there is no directional component to it. All that matters is the electric constant, k, the value of the charge, q, and the distance from that charge, q.

$$Electric\ Potential\ from\ a\ Point\ Charge: V = \frac{kq}{r}$$

Positive charges produce positive potentials or voltages, and negative charges produce negative potentials or voltages. To find the total electric potential at a given location, you simply add up the effects from each individual charge.

For any given environment, the total electric field and potential can vary depending on the exact location. Rather than displaying a string of equations to show their values at every location, a graphical representation of the fields and potential is often employed.

For the electric field, lines are drawn pointing away from the positive charges and towards the negative charges. Arrows are included to indicate their direction. As an example, for a positive point charge, the electric field lines point like rays of light away from the point charge. Where those lines appear more concentrated or closer together, the electric field has a greater magnitude.

For the electric potential, lines are drawn that indicate regions and locations where the potential has the same value (3 Volts, 5 Volts, etc.). These are called Equipotential Lines. These function as a topographical (or elevation) map for the electric effects in that environment. Incidentally, the equipotential lines always run perpendicular to the electric field lines. For a point charge, the equipotential lines form circles around the charge (say, for r = 1 m, r = 3 m, etc.). On the following pages are two examples of these electrical "maps" for various charge distributions.

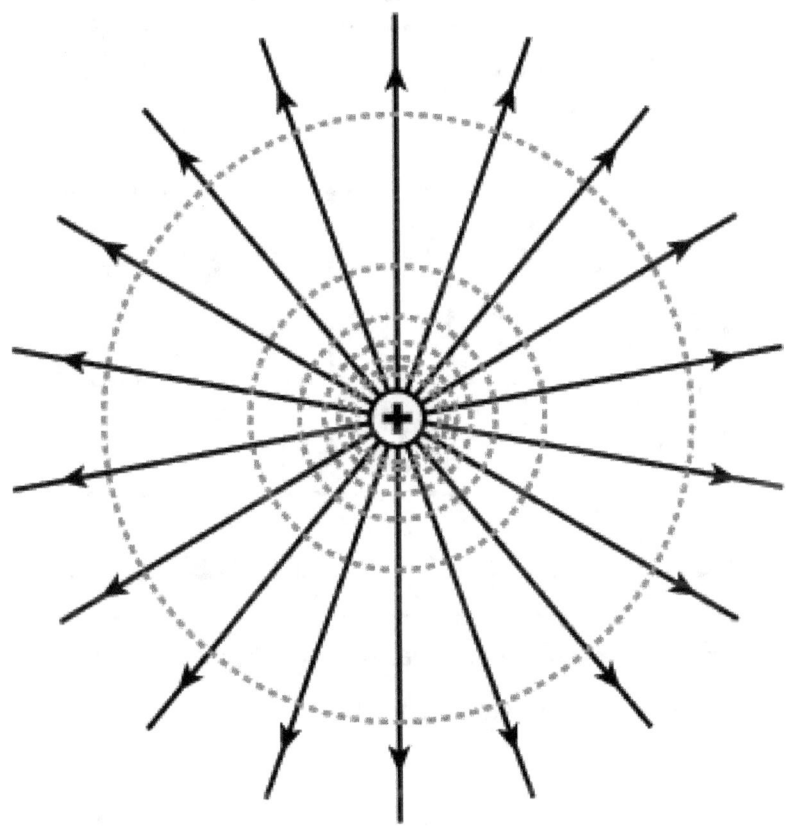

Electric Field (Arrows) and Electric Potential (Dotted Lines) for a Positive Point Charge.

Chanchocan. "Equipotenciales.PNG" (2006, August 1). Wikimedia Commons, Retrieved from *https://commons.wikimedia.org/wiki/File:Equipotenciales.PNG*

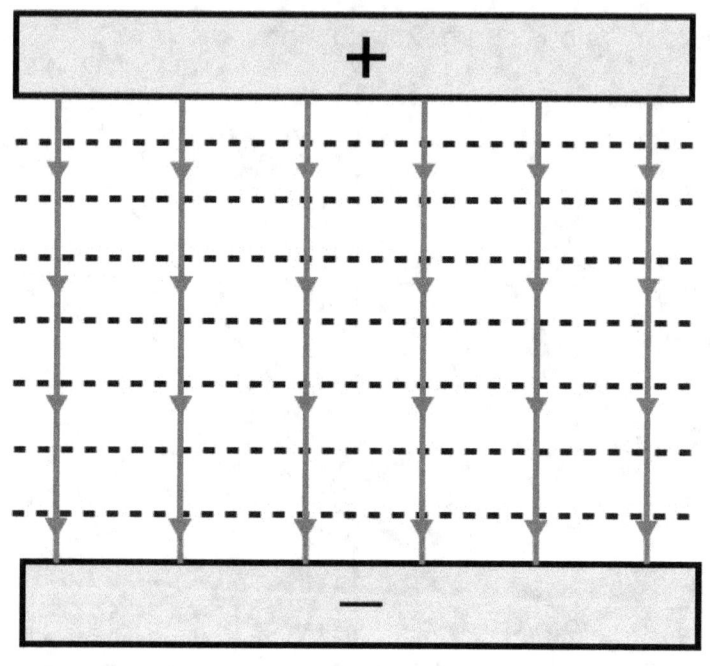

Electric Field (Arrows) and Electric Potential (Dotted Lines) for Parallel Plates of Charge.

Magnetism and the Effects of Moving Charges

Electrical Forces, Fields, and Potential all occur due to charges existing in the first place. Once those charges start to move, secondary effects appear, and that is Magnetism. A similar framework used for electrical effects is also used to examine magnetic effects. Here, the **Magnetic Field**, B, is the vector that describes all of the magnetic effects due to the moving charges in an environment. The units for the magnetic field are Teslas, T, but are commonly expressed in mT or μT. For example, Earth's magnetic field is around 40 μT. Just as electric fields cause electric forces on charges, magnetic fields cause magnetic forces on moving charges.

The typical starting point for observing magnetic effects is through bar magnets. Bar magnets always have a North Pole and South Pole, and those poles can never be isolated in a single material. If you break a bar magnet into two pieces, you end up with two bar magnets, each with their own North and South poles. Similar to electricity, like poles repel, and unlike poles attract. The strength of the magnetic force between the poles also varies with one over the distance squared. The bar magnet's field always points away from the North Pole and towards the South Pole. The magnetic field lines for a standard bar magnet are illustrated in the following figure.

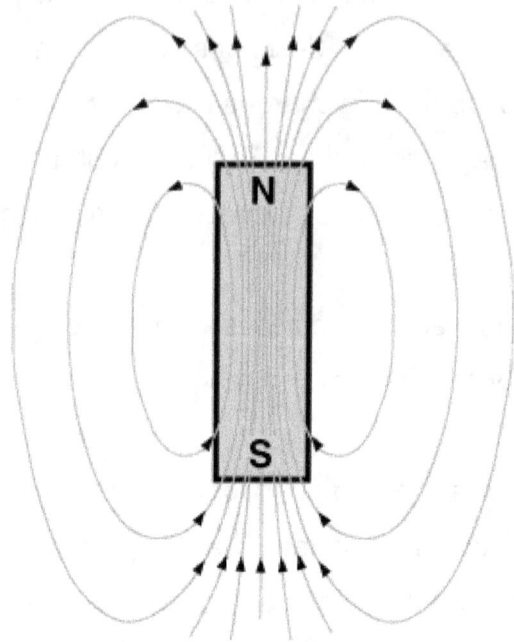

Bar Magnet's Magnetic Field Lines. Reprinted from Field Lines around Magnets.

Naik, P. Sumanth. "Barmagnet1.png" (2014, April 12). Wikimedia Commons, Retrieved from https://commons.wikimedia.org/wiki/File:Barmagnet1.png

Although bar magnets have been studied for hundreds of years, it was not apparent until the 1800s that moving charges were the ultimate source. In this case, the bar magnet's effects are due to the intrinsic spin of the atoms in a material being in alignment. In most materials, atoms have their spins in random directions. For certain metals, like iron and neodymium, their atoms can be set in a certain alignment, causing an overall magnetic effect. These are called ferromagnets. Other materials, called paramagnets, can be temporarily magnetized when exposed to other magnetic fields. Refrigerator doors are typically made out of paramagnetic materials, allowing permanent magnets to stick to them. Some materials, called diamagnets, actually arrange their atoms to oppose an external magnetic field. However, most materials, including most organics, do not significantly respond to magnetic fields.

Magnetic fields are produced by moving charges. Unfortunately, the math needed to describe the building blocks for the magnetic fields can be fairly complicated. For a moving point charge, the magnetic field is based on the value of its charge and its velocity, and also has an inverse square relationship for the distance[*]. A more important field source is that from a straight current-carrying wire. In this case, the magnetic field goes like one over the distance, without the square[†].

---

[*] The magnitude of the field from a point charge is $|B| = \left|\frac{\mu_0 q v}{4\pi r^2}\right|$ where $\mu_0 = 4\pi \cdot 10^{-7} \, T \cdot \frac{m}{A}$. When direction is included, it starts to get messy.

[†] The magnitude of the field from a long current-carrying wire is $|B| = \left|\frac{\mu_0 I}{2\pi r}\right|$

It turns out that there is always a "curved" nature to how magnetism functions[*]. For example, for this straight current-carrying wire, the magnetic field curls around it like co-centric circles. This is illustrated in the figure below.

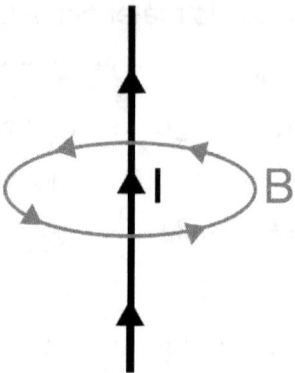

If instead the current-carrying wires are wrapped in a set of coils (a solenoid), as in the following figure, the magnetic field inside the coils is roughly constant and points along the axis of the coils. Additionally, the magnetic field outside the coils looks like that of a bar magnet!

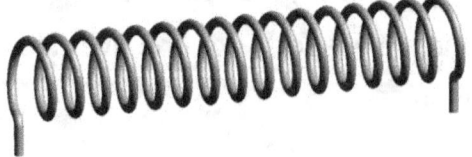

3D Illustration of a Solenoid

Zureks. "Solenoid-1.png" (2008). Wikimedia Commons, Retrieved from https://commons.wikimedia.org/wiki/File:Solenoid-1.png

---

[*] It is the result of the vector cross product being part of both the magnetic field and the magnetic force.

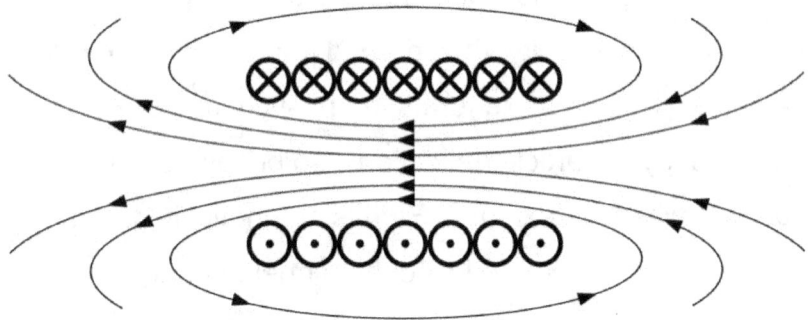

Magnetic Fields Lines for a Solenoid.

Nogueira, Nuno. "Solenoid Rotated.svg" (2009, August 3). Wikimedia Commons, Retrieved from https://commons.wikimedia.org/wiki/File:Solenoid_Rotated.svg

These magnetic fields then cause magnetic forces on other moving charges or current-carrying wires. The magnitude of the magnetic force on a moving charge is equal to the value of the charge, q, the charge's speed, v, and the magnetic field, B, at its location.

$$|F| = |qvB|$$

However, the direction of the magnetic force is quite different from what you may expect. The magnetic force points perpendicular to both the charge's velocity and the magnetic field at the charge's location. This means that moving charges do not necessarily speed up or slow down when exposed to a magnetic field – they get deflected instead*. If the magnetic field is constant and perpendicular to the charge's velocity, the charged particle ends up moving in a circle. This principle forms the basis of mass spectroscopy†. Particle deflection also becomes very important when we examine Earth's magnetic field. The sun and other sources create lots of high energy charged particles that would break apart molecules (including DNA) should they make it through the atmosphere. Instead, they hit Earth's magnetic field and get deflected and trapped in loops, keeping us safe.

Linking Electricity and Magnetism

In this chapter, we looked at how charges, by themselves, create electric fields and experiences forces

---

* In fact, magnetic fields do zero Work. Additionally, there is no scalar version of the magnetic field. The potential for magnetic fields is also a vector. In cases where objects accelerate due to some sort of electromagnetic interaction, the original current source or power supply is the starting point for energy transfer.

† Since the particle moves in a circle, we can apply our centripetal acceleration equation: $a = \frac{v^2}{r}$. Combining that with $F = ma$ and $F = qvB$ and rearranging terms, we end up with an equation describing the charge per mass ratio of a particle in terms of the particle's speed, the radius of the circular arc it travels, and the magnetic field at its location: $\frac{q}{m} = \frac{v}{rB}$. All the terms on the right-hand side of the equation can be set and measured.

from other charges and electric fields. Then, we looked at how moving charges create magnetic fields and experience forces from other moving charges and magnetic fields. We used a similar framework to discuss both of them. Furthermore, the electric and magnetic fields that you observe depend entirely on your perspective (basically, if you are moving with the charge or not).

For example, imaging your friend, Fred, is flying on an airplane, and Fiona is staying at the airport. If the airport were a charged object, Fred would see a magnetic field, since he is moving with respect to the airport. However, Fiona, who is staying still, would only see an electric field. Likewise, if the airplane were a charged object, Fiona would see a magnetic field, since it is moving relative to her, and Fred would only see an electric field, since he is moving with the airplane or is stationary with respect to it. In physics speak, we say that Electric and Magnetic Fields are completely indistinguishable depending on your frame of reference and are completely linked. This is why these two topics are normally brought together under the term, *Electromagnetism*. They are different manifestations of the same phenomenon.

We can actually go a step further by ignoring charges, for the moment, and just focus on the fields themselves. A charge, by itself, always generates an electric field. A moving charge also has a "moving" electric field. We know that a moving charge creates a magnetic field, so we could also interpret that a "moving" or changing electric field creates a magnetic field. Since what we see depends on our perspective, we should be able to swap these terms and still arrive a true statement. Therefore, a changing magnetic field should also create an electric field. This is a bit of an arcane statement, but it ends up being critically important and has a special name: **electromagnetic induction**. These previous statements are summarized in the table below[*].

Stationary Charges	Moving Charges
Experience Electric Forces from Other Electric Fields	Experience Magnetic Forces from Other Magnetic Fields
Sources of Electric Fields	**Sources of Magnetic Fields**
Stationary Charges	Moving Charges
Changing Magnetic Fields	Changing Electric Fields

The main application for electromagnetic induction is in how we get nearly all of our electric power. Let us go through a very rough model of a power plant. What electrically distinguishes one power plant from another is the first step in the process: causing a turbine to rotate any way

---

[*] The differential equations going along with the chart are called Maxwell's Equations.

possible. Wind turbines use wind, and hydroelectric dams use falling water. Most often, the turbine is rotated by heating water to steam. Coal, natural gas, and nuclear power plants have different methods to do so.

In our simple model, permanent magnets are attached to the turbine. Those permanent magnets generate a magnetic field, and once those turbines start to spin, the magnets rotate, and their magnetic fields change. Now, we utilize electromagnetic induction, where a changing magnetic field creates an electric field.

Let's say that there are some wires near the turbine. Electrons are in the atoms of those wires, and they see that induced electric field. We know that those electrons are going to experience a force due to that electric field (from $F = qE$). With that force, those electrons are going to accelerate. Moving electrons is nothing more than electrical current, which then results in our electrical power! Again, a changing magnetic field (for example, by rotating magnets) creates an electric field, which will then cause current to flow in nearby wires.

Another useful case study for electromagnetic induction and the connection between electricity and magnetism is in speakers and microphones. Speakers and microphones are electrically identical but operate in reverse. In a simple model, an amplifier is connected to some wires, and those wires are connected next to a flexible cone that has a permanent magnet attached to it*.

First, let us examine how a microphone functions. Sound waves, which are variations in pressure, cause the cone to vibrate. The wires attached to the cone are then pushed closer to the magnet, so the magnetic field at the wires' location changes. That changing magnetic field creates an electric field, which then causes current to flow through those wires. The current is the electric audio signal!

For speakers, the amplifier outputs an electric audio signal, which is a varying current. Current-carrying wires generate magnetic fields. The permanent magnet near those wires experiences a magnetic force due to that magnetic field. As the current changes with the electric audio signal, the strength of its magnetic field changes. As that magnetic field changes, the magnetic force between the magnet and the wires changes, causing the wires and the attached cone to move up and down, creating variations in pressure, which are the sound waves! Again, electricity and magnetism are completely linked.

---

* The cones are designed differently for speakers and microphones, but again, the electrical components are identical.

Ch. 8 – Circuits

Important Terms and Conventions in Circuits

The primary application of electromagnetism is in circuitry. When we examine what each circuit component does and how the circuit functions as a whole, we are primarily interested in the Electric Potential (V) and Current (I). This is because they can both be easily measured and conservation laws can be applied to both of these quantities/

The Electric Potential, as defined in the previous chapter, is the scalar that describes all of the electrical effects due to the charges that are present in an environment. Its units are Volts, V, so the terms "Electric Potential" and "Voltage" are used interchangeably. The Voltage for a component is usually in reference to the Zero Potential Point or Ground of the circuit. The ground could be the negative terminal of a battery or power supply, or the metal chassis of a device.

Current is the flow of charges over time. Its units are Amps, A, but are commonly expressed in mA. By convention, current flows from a positive voltage to a negative or zero voltage*.

Voltage sources and currents do not have to be fixed or stable. We categorize current flow as either being Direct Current (DC), where current only flows in one direction, or Alternating Current (AC), where current changes direction in time. In AC signals, the electrons do not go from one end of the circuit and back; instead, they shuffle back and forth in accordance with how the voltage varies. Both AC and DC signals can be steady or varying, but the key distinction is the DC signals only flow one direction whereas AC signals can flip. Whether a current flow is AC or DC can make a significant difference in how certain circuit components operate.

---

* This is because when electricity was first studied, it was assumed that the positive charges move. In reality, the electrons move from a negative potential to a positive potential. The convention still works though.

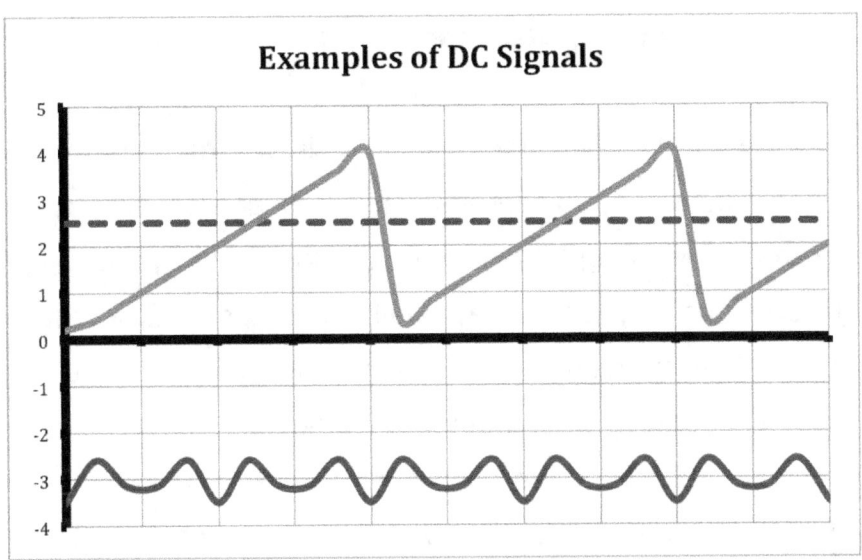

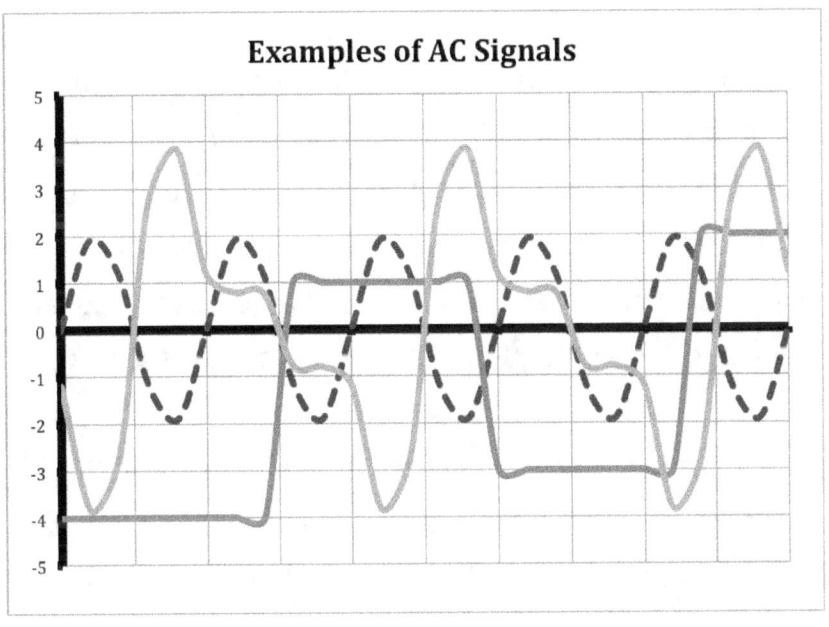

Electrical Resistance and Ohm's Law

Every conventional material resists current flow to some extent. Conductors are materials that allow electrons to easily flow through them (such as most metals), and Insulators are materials that greatly resist current flow (such as most organic compounds). Semiconductors fall in the middle, and have other special properties that will be discussed later on. Superconductors have zero resistance due to quantum effects, and exist for certain materials at very, very low temperatures.

This **electrical resistance**, R, is measured in Ohms, Ω. Typically, resistances in circuits are on the order of kΩ. Current flowing through a resistive material produces an electrical effect. This is described by Ohm's Law: the electric potential or voltage, V, due to that resistance equal to the current, I, flowing through it times its resistance, R.

$$V = IR$$

We can also look at the power associated with that current flow. The electrical power, P, flowing through that circuit element is equal to the current, I, times the voltage, V.

$$P = IV$$

Most electronic equipment and parts have limitations on what sort of current, voltages, and power they can handle. For example, speakers are typically rated in terms of their Resistance, Ω, and Power, W.

Conservation Laws for Electric Current and Potential

Again, when we look at how circuits function, we are primarily interested in what goes on with the voltage and current. Both quantities are relatively easy to measure and both have associated conservation laws, called Kirchoff's Laws.

Kirchoff's First Law is the Conservation of Charge applied to circuits. Charges cannot magically appear or disappear arbitrarily in a circuit. Even capacitors, where charges are separated, have no net charge. This conservation of charge can be applied a few different ways. First, in any junction of wires, the current flowing into that junction must equal the current flowing out.

$$\Sigma I_{In} = \Sigma I_{Out}$$

An important result of this first law concerns circuit elements that are along the same path. When they are along the same path, the circuit elements are in **series**, and they all must have the same current. The current coming from the first circuit element must equal the current flowing into the second element, and so forth. When this is applied to resistances, we find that the total or equivalent resistance of circuit elements in series is the sum of each individual resistance.

$$\textit{Resistances in Series}: R_{Total} = R_1 + R_2 + R_3 + \cdots$$

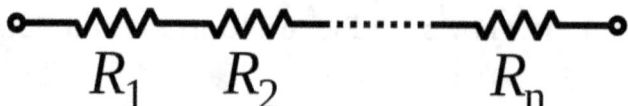

Resistors in Series

Omegatron. "Resistors in series.svg" (2007, April 29). Wikimedia Commons, Retrieved from https://commons.wikimedia.org/wiki/File:Resistors_in_series.svg

Kirchoff's Second Law is the Conservation of Energy applied to circuits. Here, we look at what happens to an electron's energy as it moves across various potential differences. Because friction does not exist at the particle level, as long as the electron ends up where it started, it can have no net change in energy. When we apply this to closed loop in a circuit, the sum of the potential differences must equal zero, or the sum of the voltages gained (through batteries and so forth) must equal the sum of the voltages lost (as the electron goes through resistors or other circuit elements).

$$\Sigma V_{Loop} = 0$$

$$Closed\ Loop\colon \Sigma V_{Gain} = \Sigma V_{Lost}$$

An important application for this is when circuit elements are along alternate paths, meaning that they are in

parallel[*]. If the starting and ending points are the same, the electron experiences the same change in energy, so their potential differences or voltages must be the same. A useful analogy is to consider the gravitational potential energy for falling objects: mgh. As long as an object experiences the same change in height, it experiences the same change in gravitational potential energy, regardless of what path it took from the top to the bottom. A similar thing is happening with an electron as it goes from one point to another along alternate paths in parallel circuits.

If we apply the voltage rule for circuit elements in parallel, we find that their combined or equivalent resistance is based on the *inverses* of each resistance. It turns out that adding resistors in parallel decreases the overall resistance of the circuit. This is because the addition of another path, regardless of its particular resistance, adds more options for the electrons to move, thus reducing the overall resistance as it moves from the starting point to the ending point.

$$\textit{Resistances in Parallel:} \quad \frac{1}{R_{Total}} = \frac{1}{R_1} + \frac{1}{R_2} + \frac{1}{R_3} + \cdots$$

---

[*] This doesn't mean geometrically in parallel. Parallel circuit elements can be arranged any which way, but they have to share the same starting and ending points.

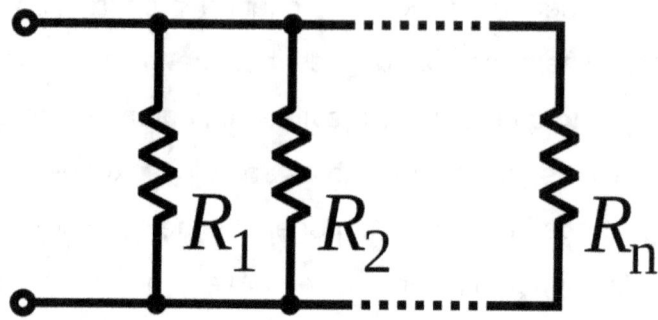

Resistors in Parallel

Omegatron. "Resistors in parallel.svg" (2007, April 29). Wikimedia Commons, Retrieved from https://commons.wikimedia.org/wiki/File:Resistors_in_paralle l.svg

Common Circuit Elements

Most every circuit is built from components from only a few categories. What differentiates one circuit from another is how those components are arranged and their number and relative values. Below are the standard symbols used for the circuit elements to be discussed.

—⌇⌇⌇— Resistor   ⊣⊢ Capacitor

⌒⌒⌒ Inductor   ⊳⊢ Diode

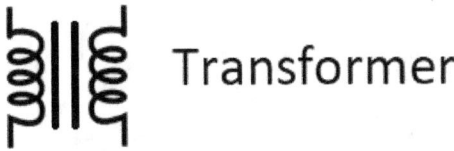 Transformer

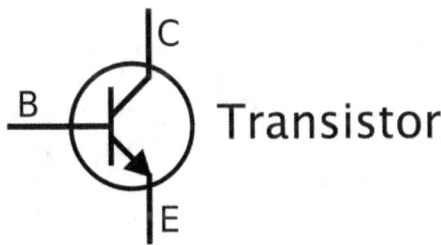 DC voltage source

AC voltage source

Transistor

Electronic Symbols for Common Circuit Elements

Jleedev. "Circuit Elements.svg" (2010, March 19). Wlkimedia Commons, Retrieved from https://commons.wikimedia.org/wiki/File:Circuit_elements.svg

Omegatron. "BJT NPN symbol.svg" (2007, September 22). Wikimedia Commons, Retrieved from https://commons.wikimedia.org/wiki/File:BJT_NPN_symbol.svg

Resistors:

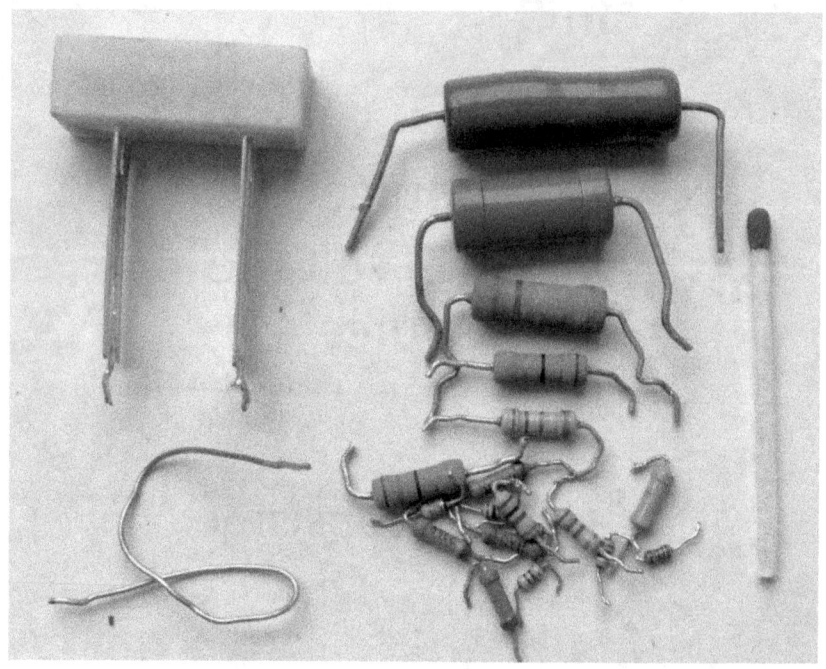

Photograph of Resistors

FDominec. "Electronic component resistors.jpg" (2007, January 4). Wikimedia Commons, Retrieved from https://commons.wikimedia.org/wiki/File:Electronic_component_resistors.jpg

The first major component in a circuit is a Resistor. All materials, barring superconductors, resist current flow to some extent. The role of a resistor is to reduce current flow through part of a circuit. Typically, they are made from metal films, carbon films, or ceramics. Electrical Resistance is measured in Ohms (Ω) with typical values in the 1 kΩ to 1

MΩ range. Resistors are rated in terms of how much power they can handle (normally less than one Watt, except for the larger ceramic resistors). Low-power rated resistors tend to be fairly small and very inexpensive. These are normally color-coded to indicate their resistance and relative tolerance with three-to-four colored bands.

Capacitors:

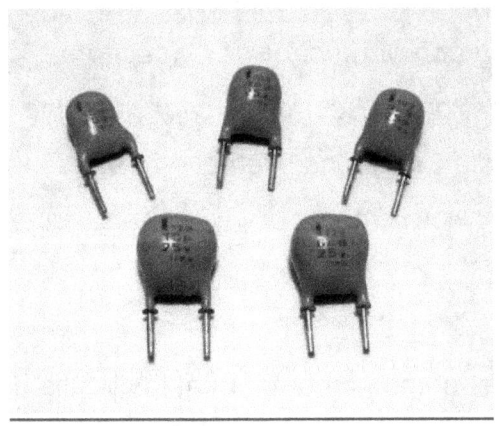

Photograph of Small Capacitors

Elcap, Jens Both "Al-Elko-SAL-Wiki-P1040340-07-02-18.jpg" (2007, February 20). Wikimedia Commons, Retrieved from https://commons.wikimedia.org/wiki/File:Al-Elko-SAL-Wiki-P1040340-07-02-18.jpg

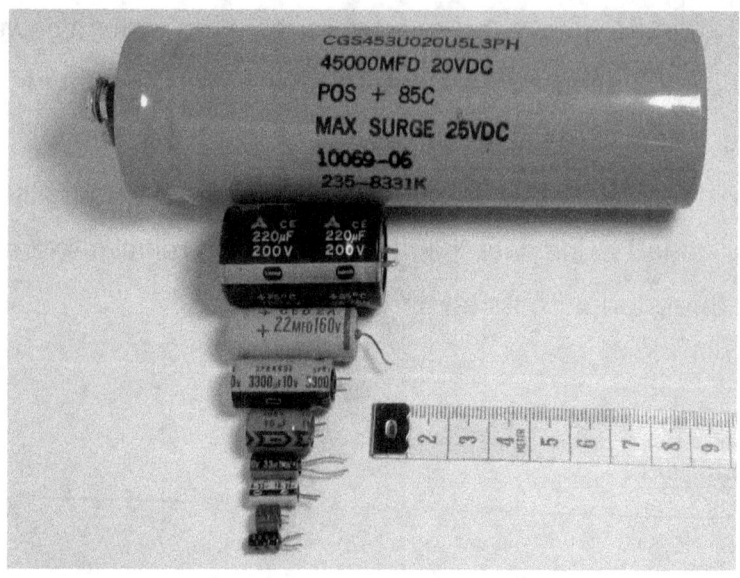

Photograph of Electrolytic Capacitors

Omegatron. "Electrolytic capacitors.jpg" (2006, June 10). Wikimedia Commons, Retrieved from https://commons.wikimedia.org/wiki/File:Electrolytic_capacitors.jpg

Another common component in circuits are Capacitors. In addition to materials resisting current flow, almost every electrical component also separates charges to some degree while still having no net charge. Capacitors are devices that are designed to separate charges so as to create a temporary potential difference that can store voltage and energy.

They can also be used to separate a varying voltage from a fixed voltage or to block certain AC frequencies.

They are normally made from metals, polymers, and ceramics. The units for capacitance, C, are in Farads, F, and are typically given in either microFarads (uF or µF) or picoFarads (1 pF = 10^{-12} F). The voltage, V, across a capacitor is equal to the separated charge, Q, divided by its capacitance, C (a property based on the capacitor's material and geometry).

$$Capacitor\ Voltage:\ V = \frac{Q}{C}$$

The energy, U, stored in the capacitor is then connected to that voltage, V, and its capacitance, C.

$$Capacitor\ Energy:\ U = \frac{1}{2}CV^2$$

Capacitors are rated in terms of how much voltage they can reliably handle, and that voltage determines their relative size. Higher voltage capacitors are normally polarized, so current and voltage must be in the specified direction. Because capacitors can retain their charge and voltage for long periods of time after being disconnected from a voltage source, you **_must_** wait a sufficiently long time for them to discharge before you can safely tinker with a device's electronics. Otherwise, *you*, rather than the circuit, will be where the current flows to reach the zero potential point!

Inductors:

Photograph of a Small Inductor.

Mike1024. "Choke electronic component Epcos 2x47mH 600mA common mode.jpg" (2007, October 11). Wikimedia Commons, Retrieved from https://commons.wikimedia.org/wiki/File:Choke_electronic_component_Epcos_2x47mH_600mA_common_mode.jpg

Inductors are the third common circuit component. Resistors are used to limit current flow, and Inductors are used to limit *changes* in current flow. This is due to electromagnetic induction. A changing current through the inductor produces a changing magnetic field. This changing magnetic field then creates an electric field and potential to oppose the changes that are occurring.

Inductors can also be used to block certain AC frequencies. The units for Inductance, which describes the inductor's capability, are Henrys, H or Hy, and are typically given in milli-Henrys. The voltage, V, across an inductor is the negative of its inductance, L (a property of the inductor based on its material and geometry), times the change in current over time.

$$Inductor\ Voltage: V = -L\frac{\Delta I}{\Delta t}$$

Inductors are rated in terms of the maximum amount of current they can safely withstand. The most common inductor is a coiled wire filled with an iron core. As such, inductors can potentially be rather sizeable and heavy.

Transformers:

Photograph of a Small Transformer

Willtron. "Small transformer, front view" (2005, October 9). Wikimedia Commons, Retrieved from https://commons.wikimedia.org/wiki/File:Small_transformer_front.jpg

Transformers are a set of two inductors placed next to each other that allow for the voltage through one inductor to affect the voltage through the other. Inductors are used to either "step up" or "step down" varying voltages. For example, a 9 V power adapter steps down the 120 V from the wall outlet to the 9 V used by the connected electronic device. Because they have two coiled wires with iron cores, transformers are always heavy and bulky.

A changing current through inductor A produces a changing magnetic field. Inductor B sees that changing magnetic field, and an electric field and potential are created through it. The ratio of the voltage through one inductor to the other is based on the number of active coils in each.

$$Transformer\ Ratio: \frac{\#\ of\ Coils\ for\ A}{\#\ of\ Coils\ for\ B} = \frac{Voltage\ across\ A}{Voltage\ across\ B}$$

Semiconductors:

The previous circuit elements can all be created using relatively conventional materials. What gave birth to modern electronics was the development of *semiconductor* devices. The name alludes to how additional energy or voltage is needed to cause these materials to conduct electricity.

Semiconductors come in two varieties: N-Type and P-Type. They earn those names by appearing to be net negative charge movers or net positive charge movers, although the "positive" charge that is moving is really an electron vacancy between atoms. Each type of semiconductor is slightly resistant to current flow, but using the two types in conjunction with each other opens a wide number of options.

Diodes:

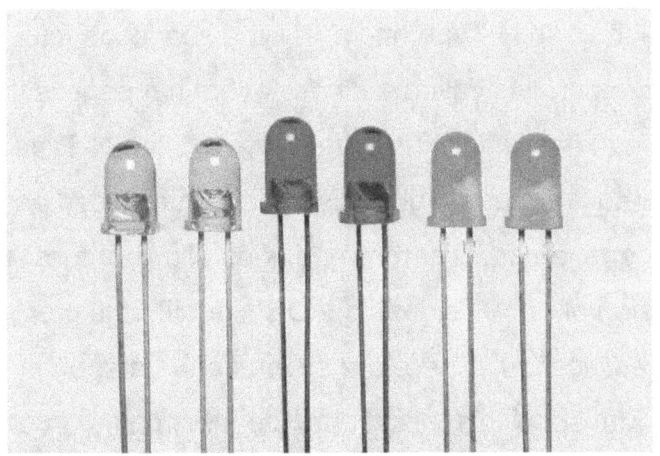

Photograph of Light-Emitting Diodes.
Afrank99. "LEDs.jpg" (2005, July 16). Wikimedia Commons, Retrieved from https://commons.wikimedia.org/wiki/File:LEDs.jpg

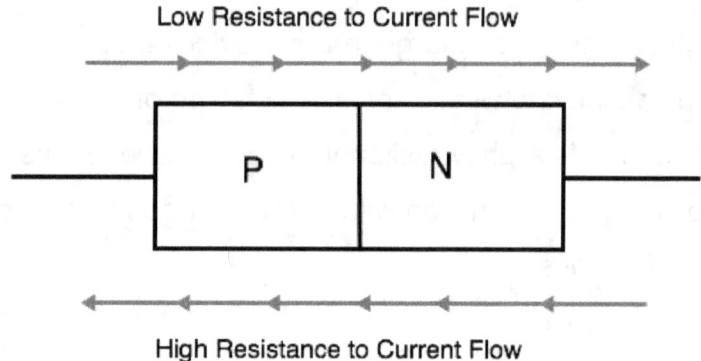

When an N-Type and a P-Type semiconductor are used together, it is called a *diode*. What happens is that current can easily flow from the P-Type to the N-Type semiconductor, but a high resistance is introduced once that current is reversed. Therefore, diodes allow for the conversion of an AC signal to a DC signal! This process is called *rectification*. There are several sub-categories of diodes with specific applications, the most common one being a Light-Emitting Diode (LED), where light is emitted when current flows the "correct" way through the diode. If wired in "reverse," diodes can be used to regulate voltage.

Transistors:

Photograph of a single Transistor.

Bomazi. "P2N2222A Transistor.jpg" (2012, December 21). Wikimedia Commons, Retrieved from https://commons.wikimedia.org/wiki/File:P2N2222A_Transistor.jpg

Transistors consist of two semiconductors of one type, and one semiconductor of the second type sandwiched between the others: NPN or PNP. One wire or lead is connected to each semiconductor. To illustrate how the transistor functions, I will use the NPN setup.

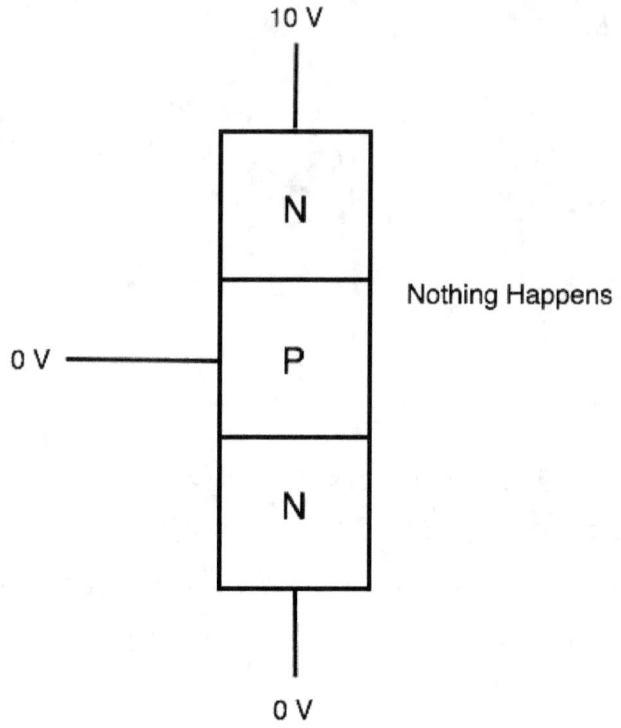

Nothing Happens

10 V is connected to the top N lead, and 0 V are connected to the other two leads. As discussed for the diode, current does not want to flow from an N to a P junction. So, under a low voltage (10 V, for the purposes of our illustration), no appreciable current will flow from the high potential to the low potential across the NP junction.

Now, let us connect 2 V to the P-type semiconductor (normally it is 0.3 V or 0.7 V). Current can easily flow from the 2 V P-type semiconductor to the 0 V N-type semiconductor. Once that happens, we can "break the rules" and have current flow from the 10 V N-type semiconductor to the P-type, then to the 0 V N-type semiconductor.

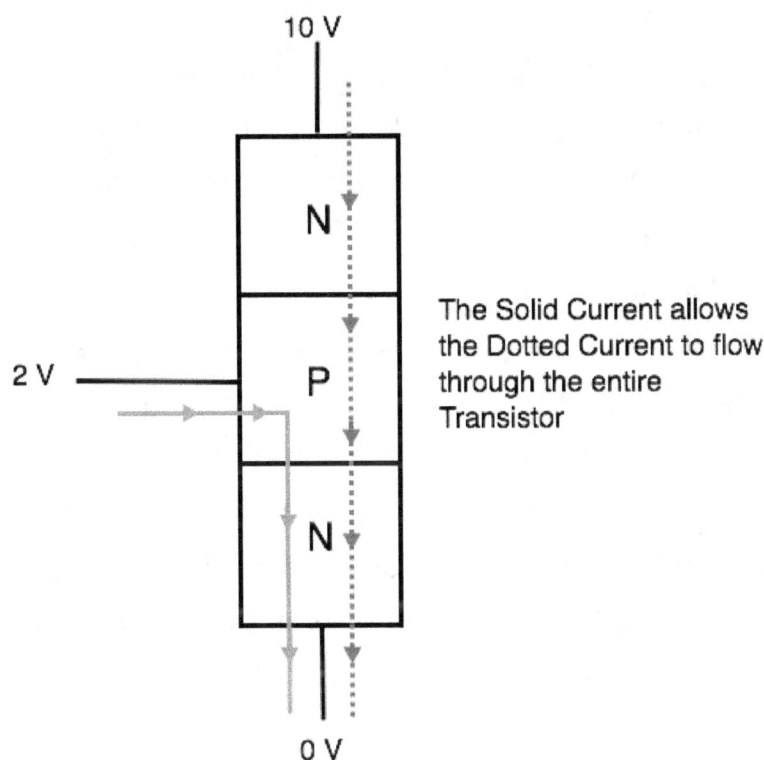

In this layout, the current flowing through part of the transistor (between the middle and end lead) can control or "turn on" the current flowing through the whole transistor (between each far end of the device). This gating or switching property allows for the development of more complex circuits. It can also be used to store information in a computer ("1" when the transistor is on and "0" when the transistor is off). Because semiconductor transistors can be built to increasingly smaller scales, very elaborate circuits can be developed over a very small amount of space. This is why the semiconductor transistor is one of the most important developments of the 20th century.

The primary disadvantage to working with semiconductors is their fragility. Most semiconductor devices cannot withstand high temperatures or voltage and current spikes. This is why heat sinks and fans are used in computers.

Interestingly enough, diodes and transistors can both be made using vacuum tubes. The natural distortion produced by vacuum tubes is still desired in audio equipment such as amplifiers. Although vacuum tubes are more durable in terms of heat and voltage, they take a great deal of space. Making a modern computer with them would cover at least the entire surface of Earth!

Ch. 9 – Optics

Light as an Electromagnetic Wave

As was discussed earlier, Electricity and Magnetism are completely linked, and are grouped under the term "Electromagnetism." It turns out that if you combine all of the base equations in defining the electric and magnetic fields, you end up with the differential form of a wave equation. Thus, Electric and Magnetic Fields travel through space as waves. Furthermore, the speed at which they travel is equal to the speed of light[*], because Electromagnetic Waves are the same thing as light! The way that one charge knows that another charge exists is through a light wave passing between them.

Let us consider some of the wave properties of light. First, light is an example of a transverse wave, where the oscillation of the electric field, the oscillation of the magnetic field, and the direction light travels are all perpendicular to each other. This means that light is inherently a 3D phenomenon. If you turn on a lamp, the light waves that are emitted in random orientations. There is a special case

---

[*] In terms of the electric and magnetic constants given earlier, the speed of light is equal to $\sqrt{\dfrac{4\pi k}{\mu_0}}$

called polarization, where a group of light waves have the same orientation (meaning that all of their electric fields are going in the same direction, and all of their magnetic fields are going in the same direction, respectively). Light can become polarized through scattering or through filtering. For example, certain sunglasses may have a polarizing filter, which only lets certain orientations of light through, thereby cutting down on the intensity of light hitting your eyes.

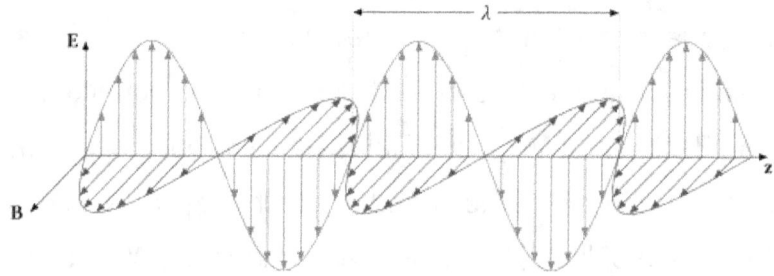

Linear Polarized Light

Francois~frwiki. "Electromagnetic wave2.svg" (2021, August 15). Wikimedia Commons, Retrieved from https://commons.wikimedia.org/wiki/File:Electromagnetic_wave2.svg

The speed of light for any individual wave is the always the same and is a fundamental constant. The **speed of light**, c, is approximately equal to $3 \cdot 10^8$ m/s. Like any other wave, the speed of light is equal to its frequency times its wavelength. Visible light only constitutes part of the electromagnetic spectrum. Going from long wavelength /

low frequency to short wavelength / high frequency, the electromagnetic spectrum is ranked as follows:

Low f, Long λ → → →   High f, Short λ
Radio Wave < Micro Wave < Infrared < Visible <
Ultraviolet < X-Ray < Gamma

The visible spectrum ranges from wavelengths of roughly 380 nm to 740 nm, or from frequencies of $(4.3 - 7.7) \cdot 10^{14}$ Hz. For reasons that we'll discuss in a later chapter, higher frequencies of light correspond to higher energies.

A very important wave property is called **diffraction**. When light passes through one or more gaps, the light waves combine and cancel out at set locations based on the wavelength of the light and the geometry of the gaps. This is due to the interference of the light waves being in phase or out of phase with each other. For example, light passing through a single circular gap does not only produce a single bright spot in center, but also produces rings of bright spots and dark spots around it, as shown in the figure below.

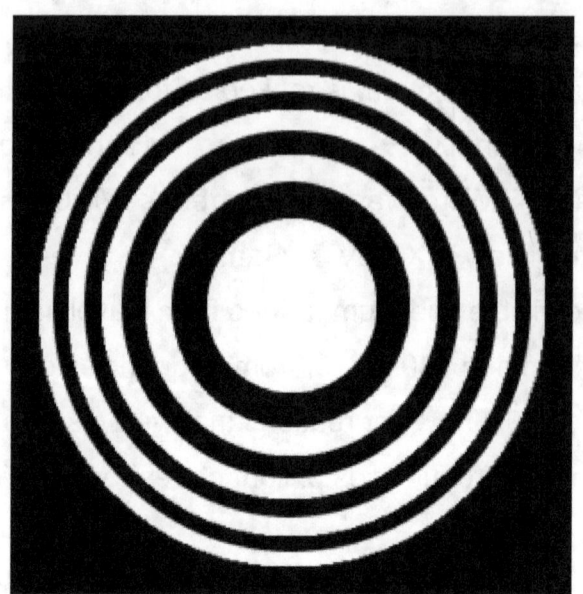

Diffraction pattern due to a Single Circular Aperture
Scudder, Laura. "Zone plate.png" (2006, March 25).
Wikimedia Commons, Retrieved from
https://commons.wikimedia.org/wiki/File:Zone_plate.png

A famous experiment on diffraction is Young's Double Slit, where light is projected onto a screen through two small, rectangular gaps. On the screen, light waves constructively and destructively interfere in a simple, predictable manner. The pattern also applies for light passing through multiple gaps. This experiment definitively demonstrated that light is a wave. The pattern of light formed and the intensities at different locations are shown in the following figures.

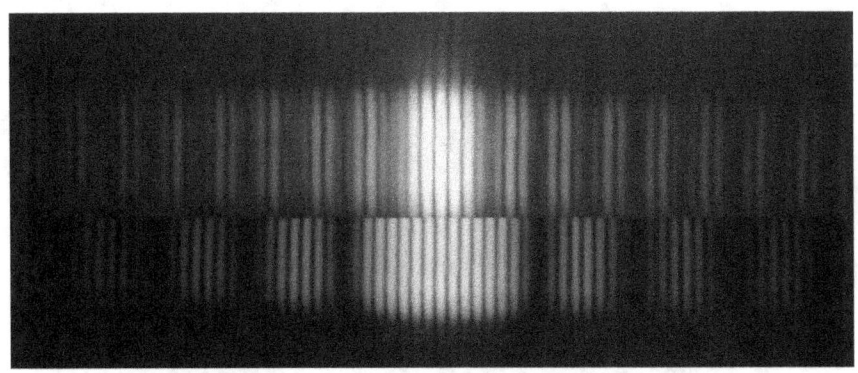

Diffraction pattern due to a Double Slit

Kuiper, Peter. "SodiumD two double slits.jpg" (2010, February 16). Wikimedia Commons, Retrieved from <u>https://commons.wikimedia.org/wiki/File:SodiumD_two_double_slits.jpg</u>

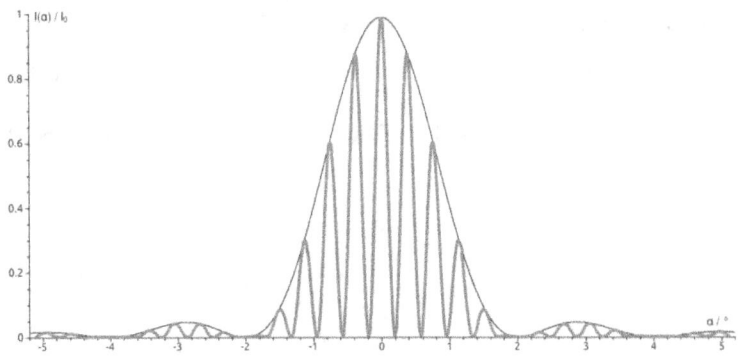

Plot of Intensity of Light that passes through a Double Slit

Keller, Klaus-Dieter "Slit double 150 28.5.svg" (2013, March 2). Wikimedia Commons, Retrieved from https://commons.wikimedia.org/wiki/File:Slit_double_150_28.5.svg

The locations of constructive and destructive interference are related to the distance between the gaps, d, the angle with respect to the line in the center between the gaps, θ, and the wavelength of light, λ.

Constructive Interference / Bright Spot:
$$\sin\theta = n\frac{\lambda}{d} \qquad n = 0, 1, 2, 3, ...$$

Destructive Interference / Dark Spot:
$$\sin\theta = \left(n + \frac{1}{2}\right) \cdot \frac{\lambda}{d} \qquad n = 0, 1, 2, 3, ...$$

This ends up being a very useful diagnostic tool for measuring very small distances. By shining light through a sample and observing the diffraction pattern, one can infer those distances. This even works at the atomic level. When X-Rays are shone through a material, the diffraction pattern that forms is based on the distance between the atoms in that material!

Light as Rays

Another way one can examine light is by treating them as rays, and seeing how the paths of those rays change under various conditions. This is normally referred to as Geometric Optics. When a ray of light hits a new material, it can either be reflected and/or transmitted.

The path for **Reflected** light follows a simple pattern:

the outgoing angle is equal to the incoming angle, or the angle of incidence is equal to the angle of reflection. It is important to note that these angles are with respect to a line perpendicular or normal to the surface of contact.

$$Relection: \theta_{Incidence} = \theta_{Reflection}$$

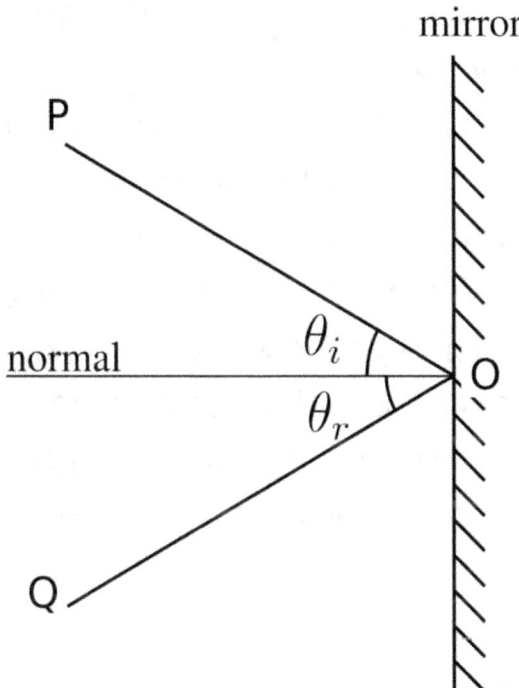

Angles for Reflected Light

Arvelius, Johan. "Reflection angles.svg" (2005, September 26). Wikimedia Commons, Retrieved from https://commons.wikimedia.org/wiki/File:Reflection_angles.svg

When a light ray is transmitted into a new material, its path can change. This behavior is known as **refraction**. This is why things appear distorted when looking down into the water. How the path of light bends is based on a property of each material, known as the **index of refraction**, n, as well as the angle it strikes the new material. For example, the index of refraction for a vacuum is 1, the index for air is about 1.0003, and the index for water at room temperature is about 1.3330. The mathematical description of this phenomenon is called Snell's Law, where the index of refraction for material A times the sine of the angle of incidence through A is equal to the index of refraction for material B times the sine of its angle transmitted through B[*].

$$Refraction: n_A \sin \theta_A = n_B \sin \theta_B$$

The following figure illustrates the phenomenon.

---

[*] Written alternatively, like in the figure below, $n_1 \sin \theta_1 = n_2 \sin \theta_2$

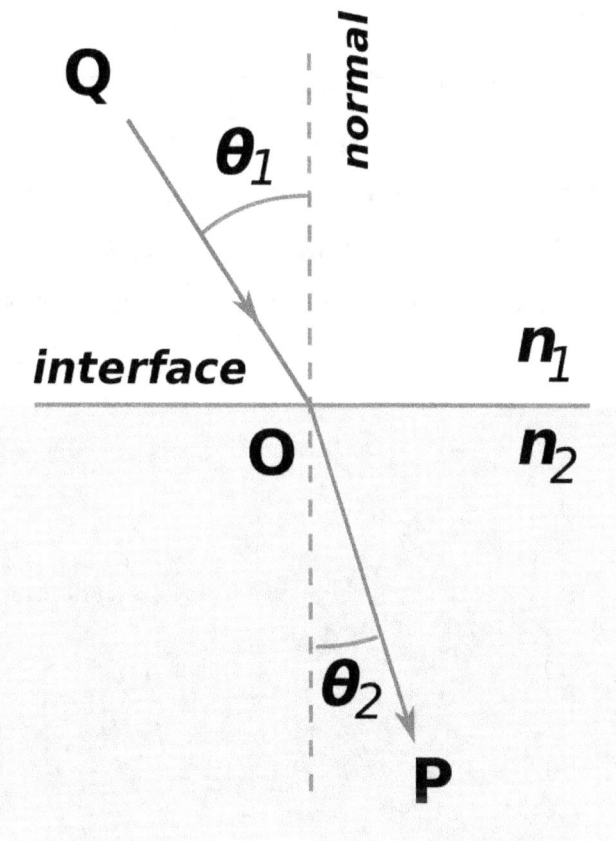

Angles for Refraction

Alexandrov, Oleg. "Snell's law3.svg" (2012, June 9). Wikimedia Commons, Retrieved from https://commons.wikimedia.org/wiki/File:Snell%27s_law3.svg

Under certain conditions, light can be prevented from transmitting through a material entirely, causing total internal reflection. This occurs when light travels at a wide angle from a high index material into a low index material*. This can happen if you are underwater looking towards the horizon. Rather than seeing the sky above you, light gets reflected off of the surface at wide angles. The figure and photograph below demonstrates this total internal reflection, where an image of a fish appears "above" the water's surface:

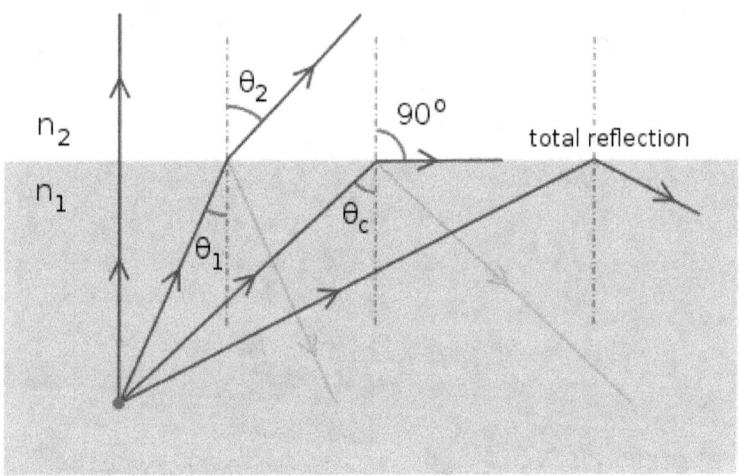

Angles for Refraction from a High Index to a Low Index material

Putland, Gavin R. "ReflexionTotal en.svg" (2019, March). Wikimedia Commons, Retrieved from https://commons.wikimedia.org/wiki/File:ReflexionTotal_en.svg

---

* Mathematically, this happens when the angle through material B is 90°, causing its sine to equal one and resulting in $\sin \theta_A = \frac{n_B}{n_A}$

Photograph of Total Internal Reflection for a fish, courtesy of Ken Shacter

This becomes an important principle when developing wave guides and fiber optic cables. By choosing a material with a high index of refraction, you can effectively "trap" light inside of it. As light goes through the material, any time it hits a boundary at a wide angle, it is reflected off of the surface and remaining inside the material.

The index of refraction also varies slightly with the light's frequency or wavelength. This is called dispersion. Usually, smaller wavelengths (or higher frequencies) of light diffract more than longer wavelengths (or lower frequencies) of light. This is why we see each the separation color in a rainbow or in prisms. The index of refraction can also vary with the light's polarization, which is called birefringence.

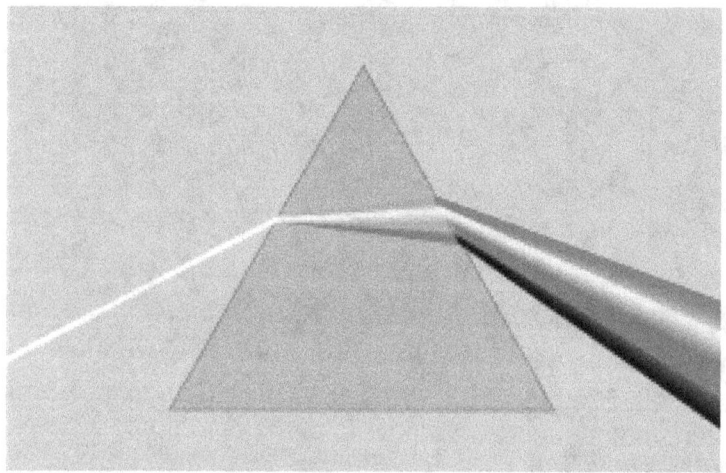

Dispersion of a Prism

Joanjoc~commonswiki. "Prism rainbow schema.png" (2005, February 11). Wikimedia Commons, Retrieved from https://commons.wikimedia.org/wiki/File:Prism_rainbow_schema.png

Lastly, the index of refraction also tells you how the apparent speed of a group of light waves decreases in a material: $v = c/n^*$. The speed of any individual light wave is still $3 \cdot 10^8$ m/s, but the light waves are constantly bouncing around the atoms in that material, being absorbed and reemitted. This makes it appear that light slows down as it travels through something.

Thin Lenses and Mirrors

The most practical application for treating light as rays is in developing lenses and mirrors. Light hits those optics at all angles, and we can apply those above rules for reflection and refraction for each of them. Fortunately, we can simplify things by just considering where the light focuses after interacting with a lens or mirror. This is described by a property of the lens or mirror called the **focal length**, f. It connects the distance from the light source to the lens or mirror (called the object distance, s) to the distance from the lens or mirror to where the light focuses (called the image distance, p) by the thin lens equation[†]:

$$\frac{1}{object\ distance} + \frac{1}{image\ distance} = \frac{1}{focal\ Length}$$

---

[*] When there is dispersion, this expression requires refinement in order to distinguish between "phase velocity" and "group velocity."

[†] In most cases, this is solved for the image distance: $p = \frac{sf}{s-f}$. Additionally, for more elaborately shaped lenses, this equation gets slightly adjusted.

$$\frac{1}{s}+\frac{1}{p}=\frac{1}{f}$$

An example of an image formed by a convex lens is given below.

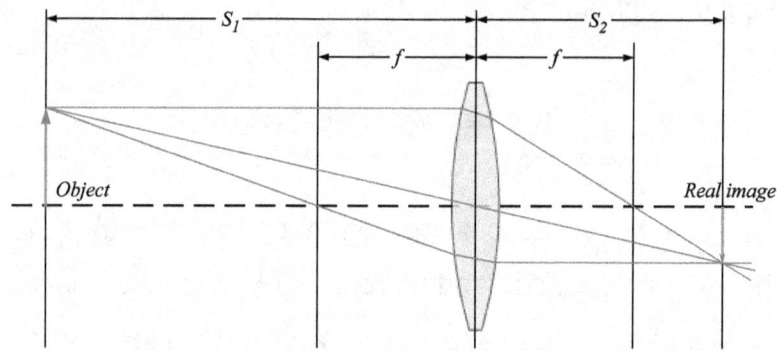

Image formed by Convex Lens

DrBob. "Lens3.svg" (2006, March 6). Wikimedia Commons, Retrieved from https://commons.wikimedia.org/wiki/File:Lens3.svg

Thin Lenses and Mirrors can be labeled as Concave (where the optic bends "inward") or Convex (where the optics bends "outward"). The following are a set of illustrations that show the path of light rays for Convex and Concave Lenses and Mirrors for light coming from infinitely far away[*].

---

[*] If the object distance is equal to infinity, the 1/s term vanishes in the thin lens equation, and the image distance becomes equal to the focal length.

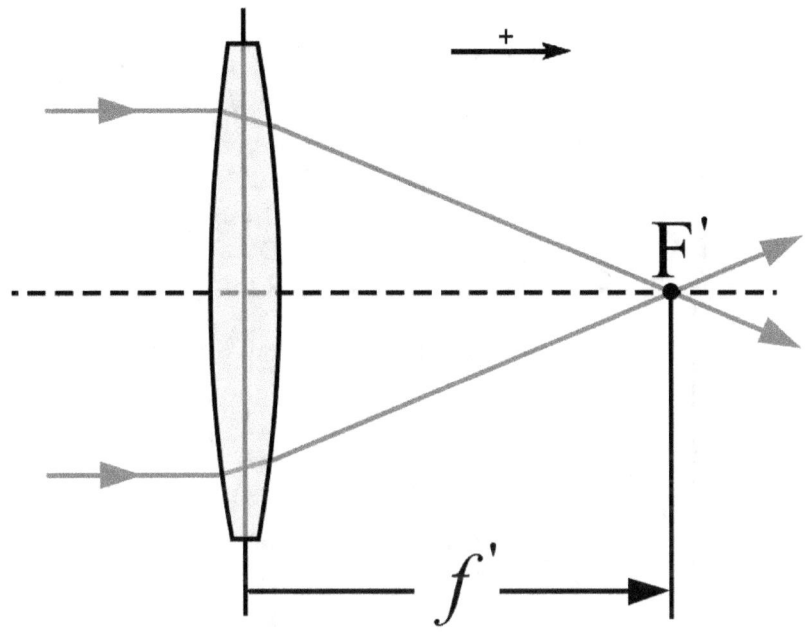

Ray tracing for a Convex Lens

JiPaul. "Focal-length-a.png" (2014, May 22). Wikimedia Commons, Retrieved from https://commons.wikimedia.org/wiki/File:Focal-length-a.png

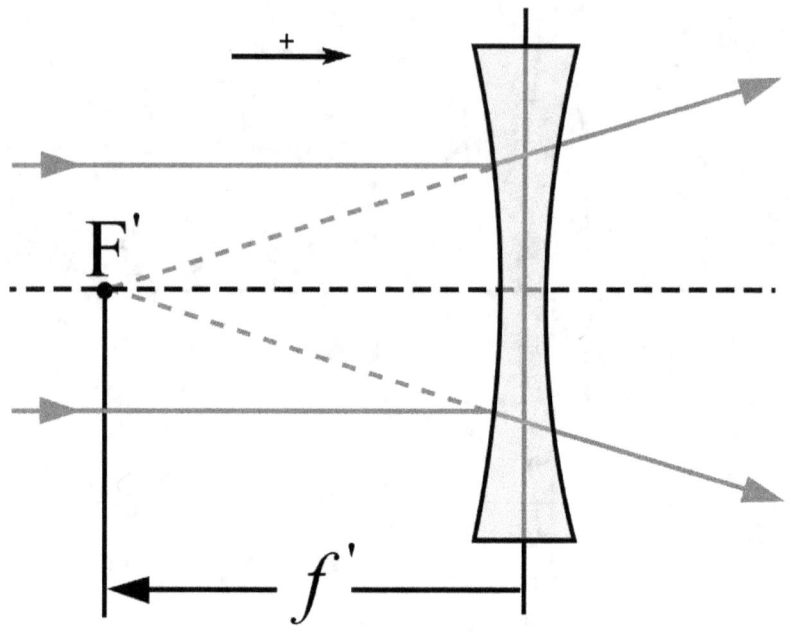

Ray tracing for a Concave Lens
JiPaul. "Focal-length-b.png" (2014, May 22).
Wikimedia Commons, Retrieved from
https://commons.wikimedia.org/wiki/File:Focal-length-b.png

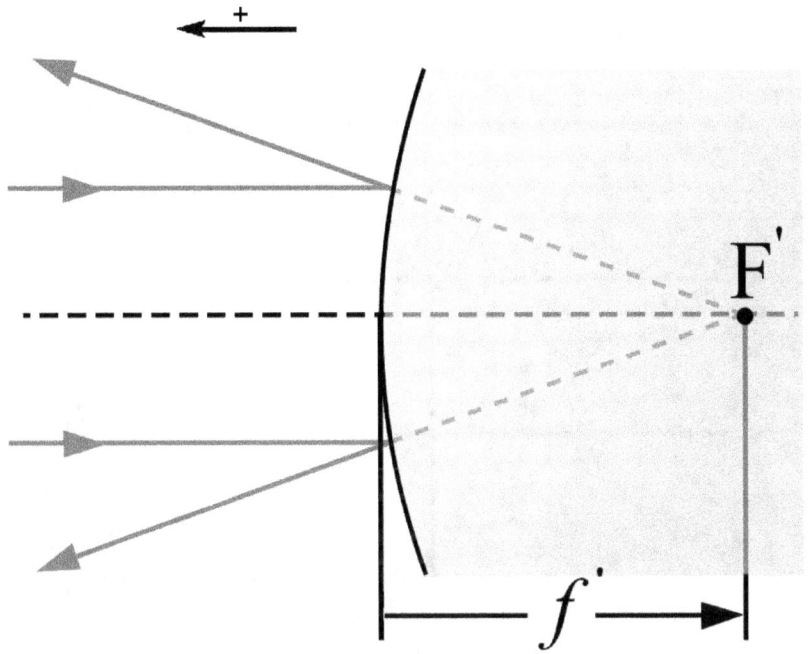

Ray tracing for a Convex Mirror

JiPaul. "Focal-length-d.png" (2014, May 22). Wikimedia Commons, Retrieved from https://commons.wikimedia.org/wiki/File:Focal-length-d.png

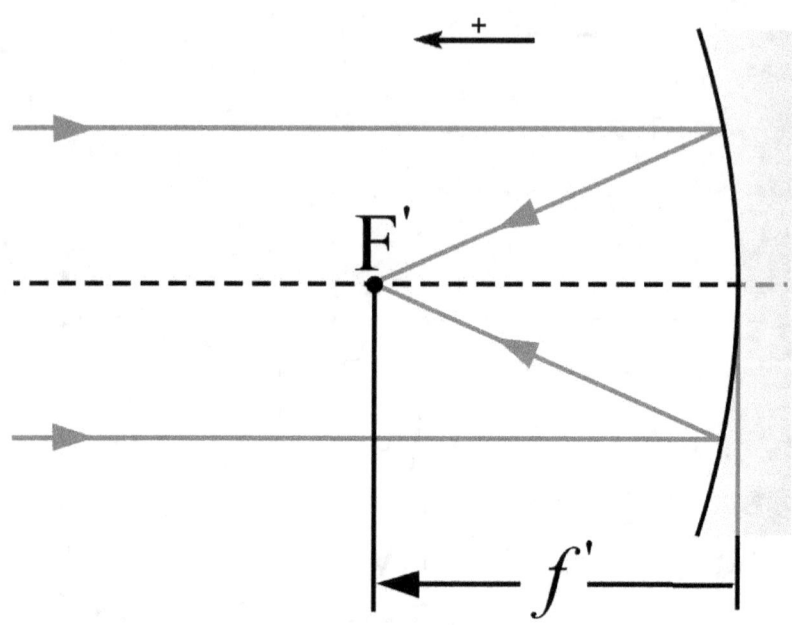

Ray tracing for a Concave Mirror
JiPaul. "Focal-length-c.png" (2014, May 22). Wikimedia Commons, Retrieved from https://commons.wikimedia.org/wiki/File:Focal-length-c.png

    The relevant equations for Lenses and Mirrors are essentially the same. Convex Lenses and Concave Mirrors both have positive focal lengths, and the Concave Lenses and Convex Mirrors have negative focal lengths.

    If an object is placed within the focal length of an optic or if the focal length is negative, the result will be a negative image distance. What happens is that the light rays diverge or split away after interacting with that lens or mirror. The light rays have to be "back-tracked" to find where the image forms. This is called a virtual image. Positive image distances mean that the light rays converge or come together after interacting with the lens or mirror and forms a real image.

The **magnification** of the image, M, is equal to the ratio of the image height to the object height. It can also be found by looking at the image distance and object distance.

$$Magnification = M = \frac{Image\ Height}{Object\ Height}$$

$$Magnification = M = -\frac{Image\ Distance}{Object\ Distance} = -\frac{p}{s}$$

Like the focal length and image distance, the magnification can be a negative number. A positive magnification means that the image is upright, and a negative magnification means that the image is inverted or upside-down. If the magnification is less than one, the image is smaller, and, likewise, if the magnification is greater than one, the image is enlarged.

In cases where there are two or more lenses or mirrors, like in a telescope or microscope, the image of the first lens becomes the object for the second lens, the image of the second lens becomes the object of the third lens, and so forth. The total magnification is the product of each lens's individual magnification.

Ch. 10 – Relativity

Special Relativity

In 1905, Albert Einstein published three key papers – each being a landmark in modern physics. The last was on Special Relativity, which connected key developments in electromagnetism to mechanics. Special Relativity can be broken down into three key elements.

The **first key element to special relativity** is that the Laws of Physics must be the same in all perspectives. In other words, it doesn't matter if you are moving or not – the rules of the game have to be the same. This may appear to be self-evident, but it is important to reaffirm. By Einstein's time, there had been several developments in reframing mechanical problems to very generic geometries, so the tools existed to describe systems from different perspectives or reference frames, and it was known that everything is moving in relation to everything else. There is no outstanding physical reason to have one "absolute" vantage point or frame of reference. The only difference is that one viewpoint may involve easier calculations versus another. The important thing is that all of the laws of physics are consistent.

The **second key element to special relativity** is that the speed of light is always constant at approximately $3 \cdot 10^8$ m/s. Measuring light's speed is different from any other speed. For example, if you are in the airport and see someone in a moving airplane throwing a ball, the speed of the ball that you measure is different than the speed of the ball that someone on the airplane would measure. This is not the case for light. If someone on the airplane turns on a flashlight, both you and that person measure that light to be travelling at $3 \cdot 10^8$ m/s, regardless of how fast the airplane is moving. The speed of light equaling $3 \cdot 10^8$ m/s comes from the electromagnetic field equations, regardless of whether the charges move or not and regardless of any particular vantage point.

Prior to the 20th century, there was a debate as to how light travelled. Every other wave travels through some medium, but what does light travel through? It was posited that light travel through some kind of "luminiferous aether," and people tried different experiments to find this substance and its nature. Furthermore, since the Earth moves through space at some speed and rotates, light, like any other wave, should measure to have different speeds in different directions. However, the aether was never detected, and interference experiments with light demonstrated that light travelled at the same speed in all directions[*]. Light always travelling at $3 \cdot 10^8$ m/s was found to be both an experimental and theoretical fact.

The **third key element to special relativity** is where the math comes in and the theory's unexpected results come from, and that is the Lorentz Transformation from electromagnetism also holds true for mechanics[†]. This describes how we mathematically go from a moving perspective to a stationary one, or vice versa.

First, let's talk about how this was originally done in mechanics using the Galilean Transformation. You are stationary at the airport, and your friend is flying in an airplane. Your friend decides to throw a ball around. You measure the ball's position, x, at various times, t. If your friend wants to track the ball's position from his perspective, x', he can use your information and incorporate the speed of the plane, v. He would use the following equation to go between your measurement of the ball's position (x) and his (x'):

$$Galilean\ Transform\ for\ Position:\ x' = x - vt$$

---

[*] See the Michelson-Morley Experiment
[†] Einstein didn't list this as an additional postulate or key element, since it is a direct consequence of the first two.

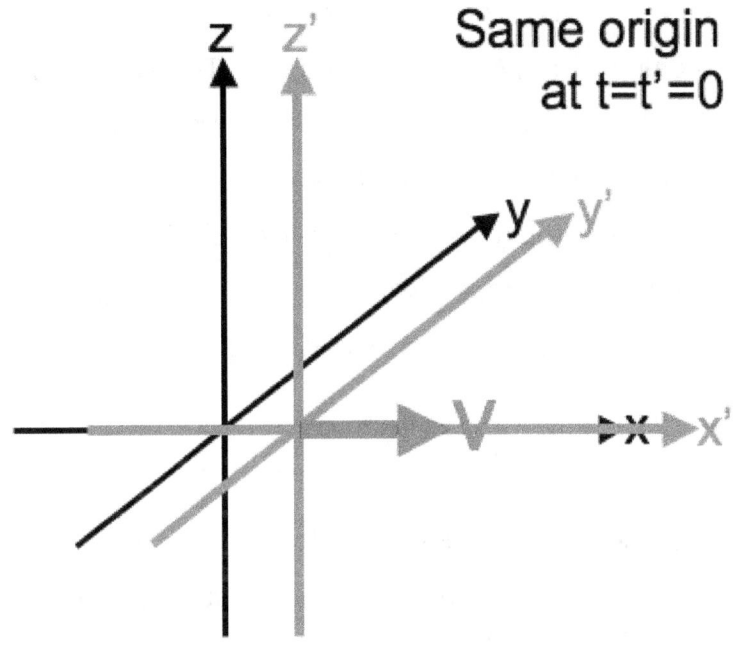

Going between stationary and moving (') perspectives
Kortemeyer, Gerd. "Standard conf.svg" (2008, May 12). Wikimedia Commons, Retrieved from https://commons.wikimedia.org/wiki/File:Standard_conf.png

Presumably, both you and your friend's clocks are functioning the same way, and measure time elapsing at the same rate, so your measured time, t, is the same as your friend's measured time, t'.

In the electromagnetism chapter, we talked about whether or not you observe the electric and/or the magnetic fields depends on your perspective or whether or not you are moving with the charge. The Galilean Transformation described above does not work for electromagnetism – it has to be adjusted, and is called the Lorentz Transformation. This also keeps the speed of light the same in all perspectives. For describing different positions, a new factor gamma, γ, is introduced, and γ is based on the speed between the two perspectives and the speed of light, c.

$$Lorentz\ Transformation\ for\ Position:\ x' = \gamma(x - vt)$$

$$\gamma = \frac{1}{\sqrt{1 - \frac{v^2}{c^2}}}$$

Initially, this doesn't seem so bad. For everyday speeds, γ is equal to one. For example, for the space shuttle travelling at 8000 m/s, $\gamma = 1 + 7 \cdot 10^{-10}$. In fact, to reach a value for γ equal to about 1.0001, the object's speed must be 1.4% of the speed of light, or $4.24 \cdot 10^6$ m/s. However, as the speed between the perspectives approaches the speed of light, γ gets huge. Furthermore, in order to keep γ a real number, nothing can go faster than light.

Where things get really strange is in how time is measured. The clocks for the stationary and moving perspectives now run at different rates! The time, t', in the moving perspective is now related to the time, t, in the stationary perspective by the following equation.

$$Lorentz\ Transformation\ for\ Time:\ t' = \gamma(t - \frac{vx}{c^2})$$

What Einstein posited is that the transformation rules for electromagnetism have to apply to mechanics as well, since the laws of physics have to be consistent. For normal speeds, the Galilean transformation is a simple and wonderful approximation, but it falls apart once things approach the speed of light.

Because of that factor γ, distances and times are different for moving perspectives versus stationary perspectives. Things that are simultaneous in one perspective are not necessarily simultaneous in other perspectives. For stationary observers, measurements for the lengths of moving objects are shorter, which is called **length contraction**.

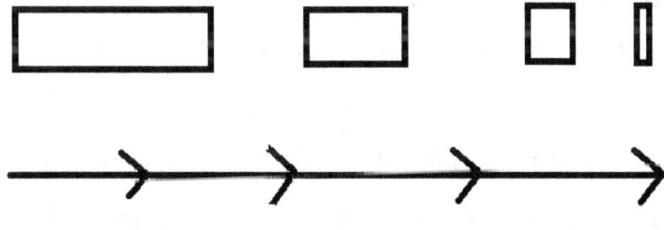

Increasing Velocity

Length Contraction Illustration

Additionally, time elapses more slowly for moving objects, which is called **time dilation**. A famous thought experiment for this is the twin paradox. One twin, Fred, stays on Earth, and the other twin, Fiona, takes off in a spaceship near the speed of light. For Fiona, 10 years passes, but for Fred on Earth, 20 years have passed.

These position and time transformations can then be applied to our base motion and energy equations. In order to make Newton's Second Law of Motion accurate for all frames of reference ($\Sigma F = \frac{\Delta p}{\Delta t}$), the momentum of an object has to include the relativistic correction: $p \rightarrow \gamma p = \gamma m_0 v$, where m_0 is the mass of the object when it is at rest (the rest-mass). An important consequence of this adjustment is that an object's resistance to force increases as its speed increase (since γ gets larger with speed), which then prevents any object from going faster than the speed of light.

When this relativistic correction for the momentum is then applied to derive the total kinetic energy of the object, we find that the total kinetic energy now equals:

$$Total\ Kinetic\ Energy = \gamma m_0 c^2$$

This is the origin of the famous $E = mc^2$, where m is the relativistic mass of the object. The phrase "relativistic mass" has fallen out of favor in recent years to better reflect the intrinsic meaning of mass and how it factors into gravity, so the Total Kinetic Energy is usually expressed as $(Total\ K)^2 = p^2 c^2 + E_0^2$, where E_0 is the energy associated with an object's rest mass.

$$E_0 = m_0 c^2$$

What that tells us is that there is always a certain amount of energy associated with an object's mass, or that mass is basically another form of energy. This fact becomes important in nuclear reactions, and some of a particle's mass gets converted into other forms of energy.

General Relativity

Special Relativity is used when changing from moving perspectives and stationary perspectives. It can be applied when the velocity is constant and during accelerations. However, gravity cannot be adequately incorporated into that theory. It took Einstein another 10 years to do so in General Relativity.

As was discussed earlier in the text, when no other forces are present, all objects fall at the same rate, regardless of mass. Furthermore, it is only when other forces are present that one can tell if gravity is there or not. We don't experience the force of gravity directly – instead, we feel the force of our bodies pressing onto the floor and such. For example, in an elevator, when that elevator first starts to move downward, we feel lighter for a moment, since the elevator is not pushing up on us as much. If we did not know better, we would not know if Earth's gravity temporarily decreased or not.

Imagine that you were at the top of a tower in a vacuum wearing a blindfold. If the floor were to suddenly disappear, you would no longer feel any forces acting on you. In fact, since you are in a vacuum, you would have no idea if you were moving at all. Furthermore, let's say that you were to do experiments in that tube and observed that the objects around you travelled in curved paths. You still would not have enough information to tell if everything was rotating or if it they were subject to an outside gravitational field.

The crux is that without outside knowledge, you have no idea if gravity is there or not, or if you were accelerating due to other forces. You can easily test if an electromagnetic field is present by using different charges and seeing what happens, but since gravity acts on all things equally, you have no means of testing its existence without the use of outside information.

General Relativity is the geometrical theory of gravity where mass and energy warp the space and time around them. The trick to understanding gravity is that it isn't really a force per se – it is an effect on the geometry of things. In the presence of a gravitational field, straight lines are no longer straight – mass and energy curve them[*]. This is akin

---

[*] The actual equations are simple in structure in terms of differential geometry. However, written out they can be very intense. Basically, a set of partial derivatives describing the

to looking at "straight" lines over the surface of Earth. We know that the Earth is spherical, so the surface isn't truly flat, even though it looks so to the naked eye. If two people who are hundreds of miles apart both start walking North, they both will reach the same point a North pole. In this sense, the surface of Earth is like a curved two-dimensional plane.

The actual geometry due to gravity is curved in both the three spatial dimensions and in time[*]. Time elapses more slowly the deeper one is in a gravitational field, just as it elapses more slowly for moving objects. This can actually be measured. Using very precise atomic clocks, it can be shown that less time passes on moving airplanes and on the space station than what time passes on the surface of Earth, due to their higher speeds and differences in gravity.

Another example of this is in muon decay. High energy charged particles from the sun create muons in

---

geometry of things is equal to a set of terms involving the mass and energy at different locations. Due to inherent symmetries, the equations can be shorted to ten unique equations for any given geometry. However, it would take a lot of space to write out those equations in a form that a person with only one year's worth of calculus would recognize, for example. Unless working in an area that requires knowledge of general relativity, most physicists don't learn the math of it in their normal course of study. If you would like to learn more about the nuts and bolts of the theory, I would highly recommend the classic text "Gravitation" by Misner, Thorne, and Wheeler.

[*] The universe has a Reimannian geometry, which is more general and has fewer postulates than Euclidean geometry.

Earth's upper atmosphere. Muons are very short-lived particles. Without incorporating relativity, those muons should not reach the surface of Earth. However, we can measure them quite easily, demonstrating that the elapsed time for those muons is less due to their fast motion and going "deeper" into Earth's gravitational field.

An important example of gravity affecting the geometry of things is in **gravitational lensing**. Even though light does not have mass, it is affected by gravity, because gravity warps the space the light travels through. This means that massive objects can bend the path of light like a lens. The figure below illustrates this.

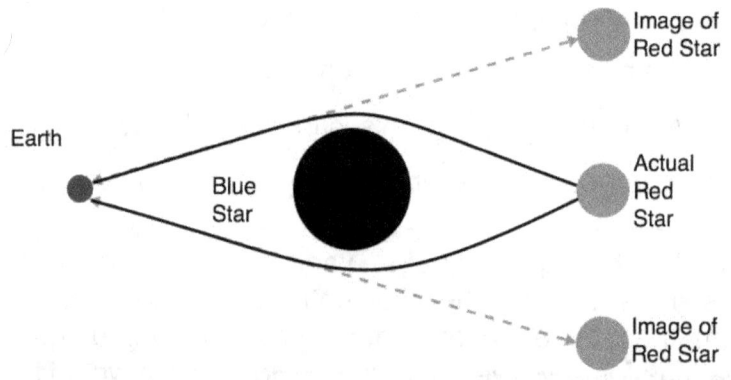

The sphere on the left side is Earth, and the sphere in the middle of the right side is a red star. A massive blue star is sitting in the middle. Ordinarily, we should not be able to see that red star. However, light emitted from that red star, when it passes near the blue star, have their paths changed due to the blue star's gravitational field, and those light rays can then reach Earth. What we see, by backtracking and drawing straight lines for the light rays, are a ring of red stars behind and around that blue star.

An early test for general relativity employed this gravitational lensing with the sun during a solar eclipse. We know the general position of the stars at night, and during a solar eclipse, the sky darkens and we can see the stars once more. However, the blocked sun is in front of some of those stars. As light passes near the sun, their paths change, and the apparent locations of those stars are different from what they were at night, thus demonstrating the gravitational lensing from general relativity.

Another important test was in describing the orbit of the planet Mercury. For all of the other planets in solar system, their orbits follow constant ellipses. However, Mercury's ellipse precesses or rotates over time, and Newton's Law of Gravity cannot fully describe why that is the case. Since Mercury is so close to the sun, the geometry of that space is more significantly warped, and general relativity can accurately describe why Mercury's orbit is different from all of the other planets.

This is all well and good, but why do these random effects matter? A very practical application for general relativity is in satellite signaling. The general relativistic effects of Earth's gravity are quite small, but they are there. If we want accurate and precise signals to and from the satellites orbiting the Earth, general relativity needs to be incorporated.

General Relativity has been hugely successful in number of areas. Even though some of its results seem strange, very little is assumed in how the theory is constructed and applied, meaning that there are few postulates in the theory as to how things work. However, it is still incomplete. There are still several cosmological effects that cannot be adequately explained. More importantly, general relativity has not been fully meshed with quantum mechanics. Special relativity is integral to quantum mechanics, but gravity cannot be described at the particle level. On the surface, quantum mechanics and general relativity are structurally very different. Quantum mechanics is like advanced set theory, and general relativity is advanced geometry. Additionally, gravity is an extremely weak force at the particle level – at least 10^{30} times weaker than electromagnetism at its best, so experiments with it are impossible at this time. One of the main goals of String Theory is bridge that gap between the two pillars of modern physics, but there are other theories and approaches being pursued[*].

---

[*] Personally, I find Gauge Theories of Gravity much more convincing and String Theory to be very dubious and intrinsically untestable. See Peter Woit's "Not Even Wrong" or Lee Smolin's "The Trouble with Physics" as to why.

For further reading on relativity that doesn't involve the messy math, I would recommend a book by the man himself: Albert Einstein's "Relativity: The Special and the General Theory."

Ch. 11 – Quantum Mechanics

Quick Note on Units for Energy

In most of Physics, we use the SI unit of Joules, J, to measure energy. In the context of atomic and particle interactions, one Joule of energy is enormous. Instead, the unit for energy that is more commonly used in this context is the **electron-volt**, eV, which represents the amount of energy required to move one electron through one volt of potential.

$$1\,eV = 1.6 \cdot 10^{-19} J$$

For the rest of the text, eV will be the most commonly used unit for energy.

Experiments and Observations Leading to Quantum Mechanics

Quantum Mechanics and Relativity are the two main pillars of modern physics. Most every field of research in physics is tied to quantum mechanics in some way, as quantum mechanics is necessary to describe how atoms and particles function. Many aspects of quantum mechanics are counter-intuitive, but it is important to note that the ideas and framework for the theory did not come out of a vacuum. There were numerous experiments and observations made in the late 19th century and early 20th century that necessitated its development. Most involved probing the nature of atoms and particles and how they interacted with light, but none had any adequate explanation using the standard approach of physics at the time. Some of these experiments are highlighted below.

Blackbody Radiation

Going back to the Thermodynamics chapter, one of the ways heat can transfer is via emitted light, called Blackbody Radiation. All objects do this, and the frequencies and intensities of light are based on an object's temperature. The figure below illustrates this for an object at a few temperatures.

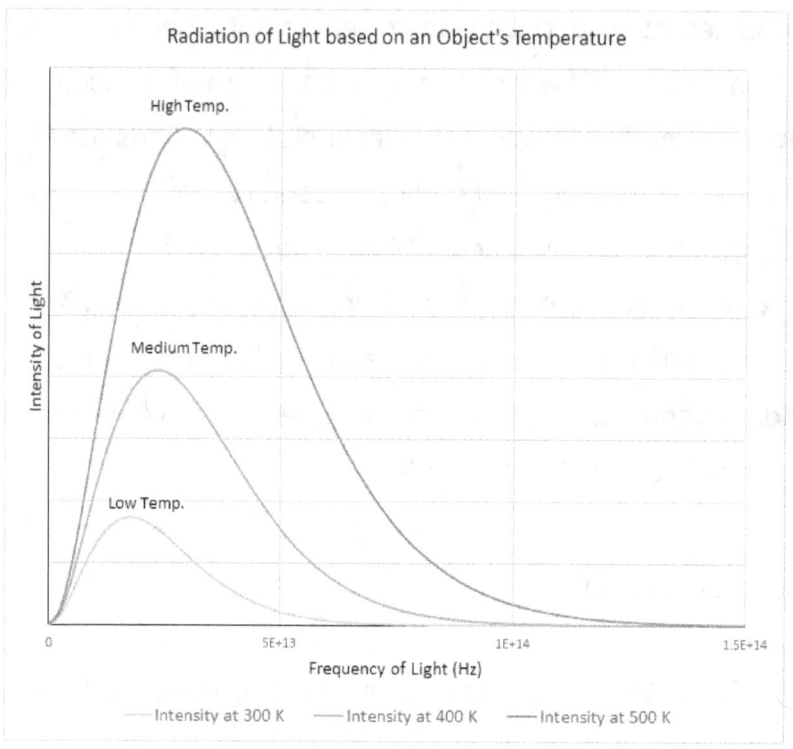

This was a very well understood and measured phenomenon. However, using the electromagnetic theories available, physicists could only theoretically reproduce the first part of the curve, and had the intensity of light constantly increasing with the frequency. This was called, with some melodrama, the "ultraviolet catastrophe," since the theory and experiment diverged starting in the ultraviolet range.

Max Planck, in 1900, was able to correctly reproduce the experimental curve by making the following main change to the existing theory. Rather than light existing at any possible energy for a given frequency, light would be restricted to certain energies. He pictured light oscillating in a cavity like a standing wave. The energy for those light waves could only exist at energies based on integer multiples of its frequency: $E = nhf$, where f was the frequency of light, n was an integer, and h was a new constant later named in his honor. Again, the drawback was that light could only have certain energies*.

The Photoelectric Effect

It turns out that when light is shone on certain metals, electrons can be emitted. However, this only happens when light reaches a certain frequency. A million infrared light bulbs cannot cause this, but one tiny ultraviolet bulb will work. This ran counter to existing electromagnetic theory, which said that the emission of electrons should be based only on the intensity of light.

In one of his 1905 papers, Einstein provided the correct theoretical basis for this phenomenon, which then earned him his Nobel Prize. The electrons needed to absorb a certain amount of energy in order to be emitted. That energy would be provided by light at given frequency: $E = $

---

*The equation that reproduces the Blackbody curve is as follows: $B(f,T) = \frac{2hf^3}{c^2} \cdot \frac{1}{e^{\frac{hf}{k_B T}} - 1}$ where B is the Intensity per solid angle and frequency, h is Planck's constant, c is the speed of light, f is the frequency, k_B is the Boltzmann constant, and T is the temperature in Kelvin.

hf, exactly like what was happening in Planck's description of Blackbody Radiation. Higher frequencies would give the electrons higher kinetic energies, and more intense light would cause more electrons to be emitted. However, the electrons always needed that threshold amount of energy and frequency of light to respond.

Compton Scattering

In the early 20th century, scientists were able to produce X-rays and see how they interacted with matter. The behaviors they observed could not be explained with the existing theories of electromagnetism. One important experiment was Compton Scattering, where is an apparent "collision" between light and electrons. Here, an X-ray is shone onto a material. The X-ray interacts with an electron, and the electron scatters off with some kinetic energy. The X-ray that is observed after the scattering ends up having a lower frequency than it did originally. The existing electromagnetic theory predicted that the frequencies should be the same.

The appropriate explanation is that the incoming light ray has a certain initial amount of energy: $E_{Initial} = hf_{Initial}$. It interacts with the electron, and gives the electron some its energy and momentum. The emitted light ray now has a reduced amount of energy, and therefore a reduced frequency: $E_{Final} = hf_{Final}$, where $E_{Electron} = E_{Initial} - E_{Final}$. Likewise, light, even though it is massless, does have some momentum corresponding to its frequency and wavelength. As some of the momentum gets transferred to the electron, the frequency of the light decreases.

Atomic Spectrum

Every element and molecule absorb and emit specific frequencies of light. This is called the Atomic Spectrum and is a very important diagnostic tool when you want to find out what makes up a certain substance. For example, below are the spectrum lines for hydrogen, helium, and iron, where the box represents the visible part of the electromagnetic spectrum.

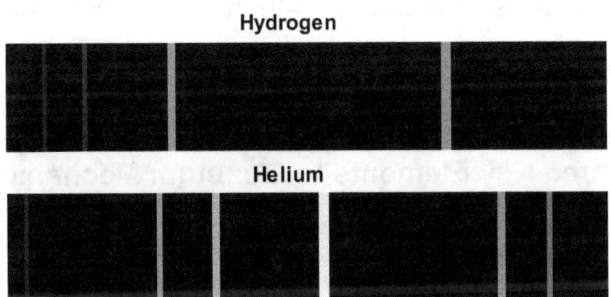

Ranjithsiji. "Atomic_emission_spectrum_of_helium.svg" (2019, July). Wikimedia Commons, Retrieved from https://commons.wikimedia.org/wiki/File:Atomic_emission_spectrum_of_helium.svg

Iron

Stiftland-Gymnasium Tirschenreuth. "Spektrum_einer_Wasserstoff-Gasentladungsr%C3%B6hre.jpg" (2018, July 17). Wikimedia Commons, Retrieved from https://commons.wikimedia.org/wiki/File:Spektrum_einer_Wasserstoff-Gasentladungsr%C3%B6hre.jpg

The patterns appear to be fairly random, and, again, there was no adequate explanation for those specific frequencies. They could easily be measured and, in some cases, have an equation to find them[*], but there was no theoretical basis in classical electromagnetism.

Three Key Elements to Quantum Mechanics

In the early 20th century, new theories for describing light and particles were developed and formalized to form Quantum Mechanics. We'll avoid any of the structural math, since it involves both partial differential equations and

---

[*] For Hydrogen, the frequencies could be found as follows: $f = 3.29 \cdot 10^{15}\ Hz \cdot \left(\frac{1}{m^2} - \frac{1}{n^2}\right)$, where n and m are integers.

advanced set or group theory. There are three main components to quantum mechanics, and they form a kind of tripod of support*. Each are tied to the others, and each represents a significant shift from classical mechanics and electromagnetism. The three key elements to quantum mechanics can be described as follows:

(1) Any object can be described as particle or as wave.

(2) Energy levels or states are discrete, not continuous.

(3) The universe is fundamentally statistical in nature.

We'll go over each of these items in greater detail.

When we examine any kind of math involving quantum mechanics, **Planck's constant**, h, shows up. Planck's constant can be expressed in terms of Joules or electron-volts: $h = 6.626 \cdot 10^{-34} J \cdot s = 4.136 \cdot 10^{-15} eV \cdot s$. It represents the smallest resolution or level of detail in the universe. It is important to note that it is a very, very small number, which is why we have to start at the particle level to immediately observe quantum effects.

---

* See David Bohm's classic text "Quantum Theory" for further details. The math is fairly intense, but the explanation for the historical and theoretical bases for quantum mechanics is excellent and approachable. A more modern and excellent mathematical treatment is Peter Woit's "Quantum Theory, Groups, and Representations."

First, let us consider how any object can be described as a particle or as a wave. Although this is a bit of a strange statement, it is not so big of a stretch mathematically. This is because any mathematical function can be described as a combination of wave functions. If we have some function to describe a particle, we should be able to do so in terms of waves[*]. In fact, we can use a single wavelength, λ, to describe a particle, and we do so using Planck's constant, h, and the momentum, p, of the particle. This is called the de Broglie wavelength:

$$de\ Broglie\ Wavelength:\ \lambda = \frac{h}{p}$$

We can easily measure the wavelengths of electrons, protons, and smaller atoms. For example, we can do diffraction and interference experiments with them, just as what was done for light back in the Optics chapter. The larger the object is, the smaller the wavelength. Given how small Planck's constant is, it is not feasible to measure the wavelength of things much larger than a molecule.

We can apply this principle to light as well. Even though light doesn't have mass, it does have momentum, since we can also treat it as a particle. This allows us to do collision experiments between light and electrons. A light particle can impart some of its momentum to an electron,

---

[*]The non-relativistic form of this is Schrödinger's Equation

resulting in the light's momentum decreasing, its wavelength increasing, and its frequency decreasing, giving us Compton scattering.

The second key element to quantum mechanics is that energy levels are discrete, not continuous. Prior to this chapter, we have had no real restriction as to the value of energy. In quantum mechanics, the energy for a particle or a system can only take on certain specific values – they are always connected to Planck's constant and one or more integers. For example, the energy for a single light wave or particle is equal to Planck's constant times its frequency:

Energy for a Single Light Wave: $E = hf$

We say that the energy from light comes in "packets," where each individual wave or particle contributes that amount of energy, hf. In the photoelectric effect, each light wave of a sufficient frequency gives the electron enough kinetic energy to free itself from the atom, and more light waves at those frequencies cause more electrons to be emitted.

But what about other particles? Going back to when energy was first discussed, the total energy for a particle is equal to its kinetic energy plus its potential energy[*]. If we apply the wave nature of a particle to that equation, we end up with integer relationships for the energy, just as what

---

[*] Friction is a macroscopic quantity, so it's not present at the particle level.

happened for light in Blackbody radiation. For example, the first-order energy for electrons in Hydrogen is directly tied to a single integer, n.

$$E_n = \frac{-13.6 \ eV}{n^2}, \quad n = 1, 2, 3, ...$$

The -13.6 eV comes from Planck's constant, the electric constant (k, from the Electromagnetism chapter), the mass of the electron, and the charge per proton or electron, e.

The last key element to quantum mechanics is that the universe is fundamentally statistical in nature. Going back to Newton's Laws, if you knew all of the forces in a system and all of the particles initial positions and velocities, you could figure out what that system would do from now until the end of time. In quantum mechanics, we are fundamentally limited in what we can know about a system. We can determine the possible outcomes and their respective probabilities, but we cannot know those outcomes ahead of time. It is like when you flip a coin, you know that you have a 50% chance of it landing on heads and a 50% chance of it landing on tails, but you don't know which one it will be in a particular instance. That is how the universe works on a base level.

When we apply the wavelike nature of a particle to the energy equation, we end up with a probability wave of sorts, and that probability wave can be used to find expected positions, momentums, energies, and so forth. Furthermore, some of those variables are linked, where the more you know about one variable the less you know about the other, and you cannot know either quantity exactly to whatever precision you want. This is called the Uncertainty Principle. The most common application of the uncertainty principle is that you cannot know the exact position or the exact momentum of a particle at the same time. That knowledge is limited by Planck's constant. In equation form, it looks like the following[*]:

$$Uncertainty\ in\ Postion \cdot Uncertainty\ in\ Momentum \geq Planck's\ Constant$$

Again, it is not the we know nothing about what happens. We can determine expected outcomes and measure variables to a high degree of precision (due to Planck's constant being so small), but we cannot know certain things exactly.

Electrons in Atoms and the Atomic Spectrum

Let us turn our attention to an important case study of quantum mechanics: electrons in the atom. What we'll do is apply the various key elements and results outlined above to this system. First, we say that energy levels are discrete, not continuous. This means that our electrons can only exist

---

[*] Technically, it's Planck's Constant divided by 4π

in specific energy states in the atom. Furthermore, due to the wave nature of the electron and the uncertainty principle, those electrons exist in general regions for each energy state. Despite the classic image of the atom, electrons do not orbit the nucleus like planets orbit the sun. Instead, the electrons exist in general "clouds" in different regions of the atom. This, along with the fact that two electrons cannot exist in the same state is what gives rise to electron orbitals, as illustrated in the figures below.

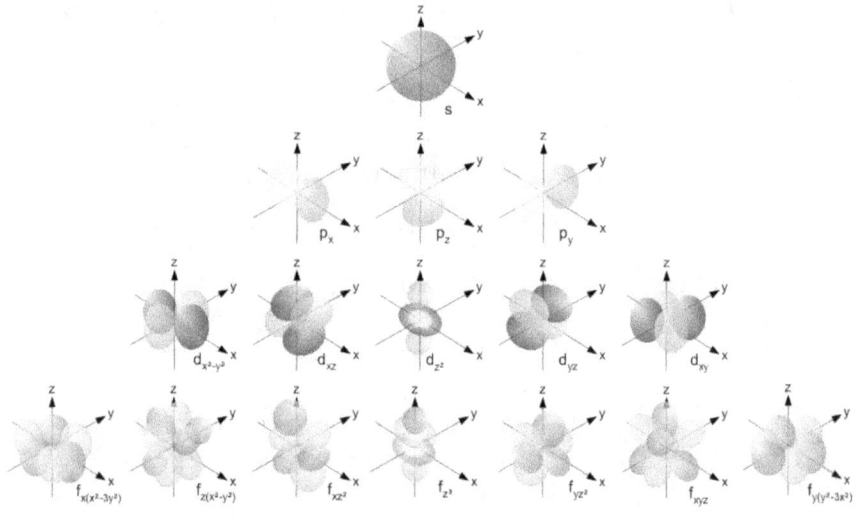

haade. "Single_electron_orbitals.jpg" (2006, November 18). Wikimedia Commons, Retrieved from https://commons.wikimedia.org/wiki/File:Single_electron_orbitals.jpg

Since our electrons exist at specific energies in the atom, they require specific energies in order to change their states. Those specific energies are provided by light waves at specific frequencies, via $E = hf$. Different atoms and molecules have different energies for their electrons. Therefore, they absorb and emit different frequencies of light, which then gives rise to the atomic spectrum described earlier in the chapter!

For example, the energy for electrons in hydrogen was given to be the following:

$$E_n = \frac{-13.6 \ eV}{n^2}, \quad n = 1, 2, 3, \ldots$$

If we wanted to go from one state, n, to another state, m, the energy required would be:

$$\Delta E = E_m - E_n = \frac{-13.6 \ eV}{m^2} - \frac{-13.6 \ eV}{n^2}$$

$$= -13.6 \ eV \cdot \left(\frac{1}{m^2} - \frac{1}{n^2}\right)$$

Let's say we wanted to go from the ground state (n = 1) to the first exited state (m = 2). The energy needed to make that transition would then be:

$$\Delta E_{1 \to 2} = E_2 - E_1 = -13.6 \ eV \cdot \left(\frac{1}{2^2} - \frac{1}{1^2}\right) = 10.2 \ eV$$

With that energy, we can use the energy of light equation to figure out the frequency of light needed to make that transition. Here, we need to use the eV·s version of Planck's constant since the energy is given in terms of eV:

$$E = hf \rightarrow f = \frac{E}{h} = \frac{10.2 \ eV}{4.136 \cdot 10^{-15} \ eV \cdot s} = 2.47 \cdot 10^{15} \ Hz$$

This would be in the ultraviolet range. We could then figure out the momentum of that light wave by using the speed of light equation ($c = f\lambda$) and the de Broglie wavelength equation ($\lambda = \frac{h}{p}$). To get our momentum in units of kg•m/s, we'll use the J•s version of Planck's constant. Rearranging and combing terms, we have:

$$c = f\lambda \;\rightarrow\; \lambda = \frac{c}{f} = \frac{3 \cdot 10^8 \frac{m}{s}}{2.47 \cdot 10^{15} \, Hz} = 1.21 \cdot 10^{-7} m$$

$$\lambda = \frac{h}{p} \;\rightarrow\; p = \frac{h}{\lambda} = \frac{6.626 \cdot 10^{-34} \, J \cdot s}{1.21 \cdot 10^{-7} \, m} = 5.48 \cdot 10^{-27} \, kg \cdot \frac{m}{s}$$

As a parting note, because Planck's constant is so small, you are only going to notice individual quantum effects, such as the wavelike nature of an object, at the particle level. However, quantum mechanics is absolutely necessary to describe how things work in the atomic level – classical mechanics and electromagnetism fall apart. Once those atoms are accurately described, this allows for the accurate description of how a materials function as a whole. All of chemistry is basically applied quantum mechanics, as it deals with interaction between electrons, nuclei, and light. Photosynthesis is the interaction between light and electrons. Electrical conduction is again how electrons function in atoms. Lasers are so prevalent in modern technology, and their development came from utilizing the discrete energy transitions for electrons and the emission of light in the process.

Ch. 12 – Nuclear and Particle Physics

Defining the Nucleus

Last chapter, we looked at quantum mechanics and how it applied to light and electrons. Now, we will shift our focus to the nucleus and other subatomic particles and apply the rules of quantum mechanics to them.

The nucleus is made up of protons and neutrons. The protons are positively charged, and the neutrons have no net charge. If only the electromagnetic force were present, the nucleus would fall apart.

However, protons and neutrons are actually composite particles, each made up of three quarks. A proton is two up quarks and a down quark, and the neutron is two down quarks and one up quark. Those quarks interact and bond together via the strong nuclear force, which can then hold the nucleus together. The strong nuclear force will be revisited later in the chapter.

The number of protons in the nucleus defines each element. What makes a helium atom a helium atom is due to it having two protons in the nucleus, carbon is defined as having six protons in the nucleus, and so forth. The number of protons in the nucleus is also called the atomic number, Z.

The number of neutrons, N, can vary for each element. As long as a there are six protons in the nucleus, it is considered a carbon atom, regardless of whether there are six, seven, eight, or however many neutrons. The total number of protons and neutrons in the nucleus is called the mass number, A, so $A = Z + N$.

Because the number of neutrons can vary for any given element, we introduce the term **Isotope** to refer to specific kinds of atoms of a given element that have a certain number of neutrons. There a few ways we denote isotopes. First, you may see the element name or symbol with a dash and the mass number, such as Carbon-14 or C-14. Another way is where the element symbol is given, the mass number is given in the top left, and the atomic number is given in the lower left (the atomic number sometimes is omitted). For example, that Carbon-14 isotope would be written as: $^{14}_{6}C$. This tells the reader that this particular isotope of carbon, which is defined by having 6 protons, has a mass number of 14, meaning that it has $14 - 6 = 8$ neutrons.

Radioactive Decay

Types of Radioactive Decay

Generally speaking, all isotopes of a given element are chemically identical. However, not all isotopes are stable. They have a high potential energy and can reduce it by emitting different particles. This is called **Radioactive Decay**. The total energy of the system is still constant – it's just that some of its potential energy from its unstable arrangement is converted to kinetic energy for the emitted particles. In fact, as predicted by special relativity, some of the isotope's mass is converted to other forms of energy!

Radioactive decay can usually be categorized into alpha, beta, or gamma decay. Alpha decay typically occurs for the heavier isotopes. Here, the isotope emits an alpha particle or helium nucleus (two protons and two neutrons) in order to reduce its potential energy[*]. For example, the alpha decay process for a Uranium-238 isotope is as follows:

$$^{238}_{92}U \rightarrow {}^{4}_{2}He + {}^{234}_{90}Th$$

The Uranium-238 isotope emits that helium nucleus, thus losing two protons and two neutrons. Its mass number

---

[*] Alpha decay is the most common example of a Cluster Decay, where a heavy nucleus splits into two nuclei. For example, Barium-114 can decay into Tin-102 and Carbon-12.

goes from 238 to 234. Having lost two protons, its atomic number goes from 92 to 90, and the uranium isotope becomes a thorium isotope!

A common application for alpha emitters is Americium-241 in smoke detectors. The Americium isotope emits those alpha particles that will then ionize the air and current is conducted through it. If smoke enters the area, the current is reduced, and the smoke detector activates.

Beta decay comes in two varieties: beta-positive and beta-negative. Both involve the weak nuclear force, a new kind of interaction. Here, a quark changes its type, and a proton becomes a neutron or a neutron becomes a proton. In a beta-negative decay, a neutron in the nucleus becomes a proton that stays in the nucleus, and an electron and an anti-neutrino are emitted. Neutrinos are a very small, chargeless particle that barely interact with other particles. The following is an example of a beta-negative decay process for Carbon-14.

$$^{14}_{6}C \rightarrow ^{14}_{7}N + e^- + \bar{v}$$

Since a neutron in the nucleus becomes a proton, the mass number doesn't change, but the atomic number increases from 6 to 7, and the carbon isotope is now a nitrogen isotope.

The stranger process is beta-positive decay. Here, a proton becomes a neutron that stays in the nucleus, and a positron and a neutrino are emitted*. The positron is the anti-particle of an electron. The following is an example of beta-positive decay for Copper-64.

$$^{64}_{29}Cu \rightarrow ^{64}_{28}N + e^+ + \nu$$

Again, since the proton becomes a neutron, the mass number doesn't change, but the atomic number decreases from 29 to 28, and the copper isotope becomes a nickel isotope.

The positron is the anti-particle of an electron. As soon as that emitted positron interacts with an electron, they will annihilate each other and two high energy light rays or gamma rays are emitted opposite of each other. This can be very useful for medical tracers. For example, ff a detector is placed around a person or object that has a beta-negative isotope injected, that detector will pick up those secondary gamma rays. Based on when and where those gamma rays are detected, a 3D image of the person or sample can be developed. This is the basis of the PET scan – Positron Emission Tomography!

The last kind of radioactive decay is gamma decay. Here, the nucleus itself is in an excited state, and emits a high energy light ray or gamma ray to reduce its potential energy. The nucleus itself does not change in its composition, so it stays as the same isotope. Last chapter,

---

* This can also happen if the nucleus "captures" an external electron.

we looked at the energy levels for electrons in hydrogen. When an electron goes from the n = 1 state to a higher energy, we say that the electron is in an excited state. A similar thing can occur for the nucleus itself. For example, here is a gamma decay reaction for a Cobalt-60 isotope.

$$^{60}_{27}Co^* \rightarrow {}^{60}_{27}Co + \gamma$$

Here, the excited nucleus is marked with an asterisk, and the high energy light ray is marked with a gamma symbol. These gamma rays can be used to kill harmful bacteria or as detectors.

If you want to find out if a particular isotope is unstable or not, a nuclide chart is used. It lists all the common isotopes, and, if they are unstable, marks the most likely decay option. The straight line is Z = N, or where the number of protons equals the number of neutrons.

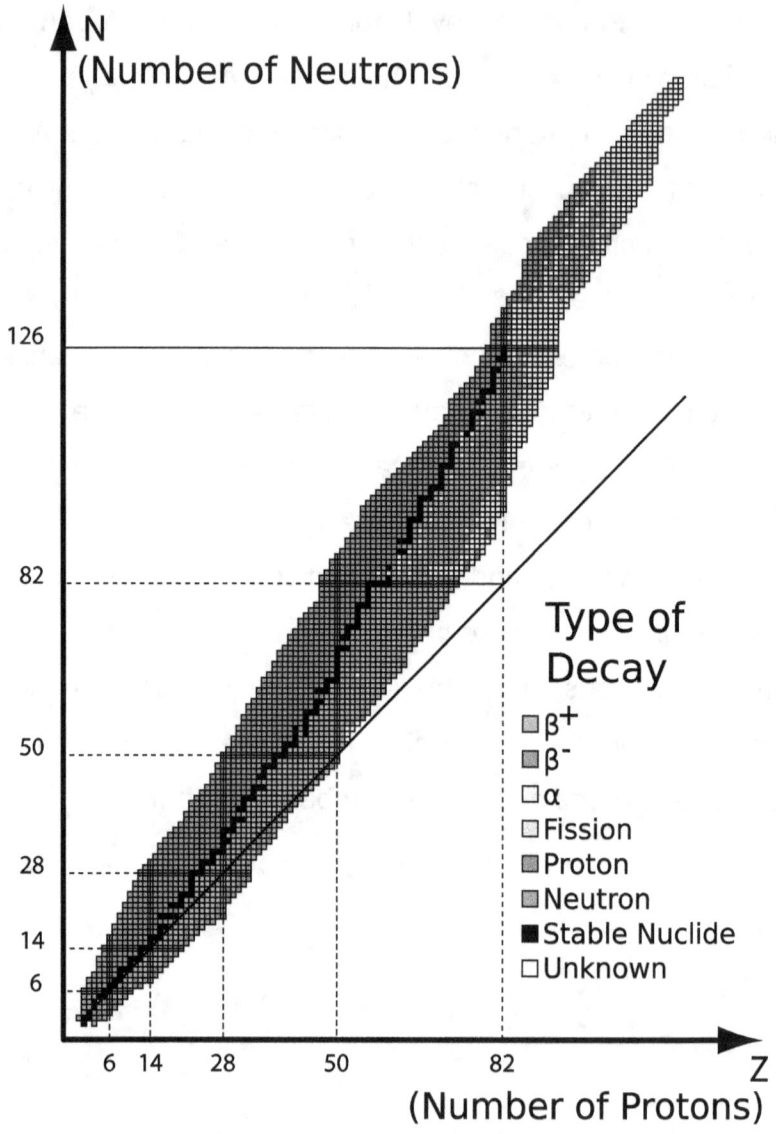

Napy1kenobi. "Table_istopes.svg" (2009, May 5). Wikimedia Commons, Retrieved from https://commons.wikimedia.org/wiki/File:Table_isotopes_en.svg

In the previous chart, you can see that as the number of protons increases, more and more neutrons are required to keep the isotope stable. Past a certain point (Lead with 82 protons), there are no more stable isotopes.

Measuring Decay Rates

As discussed in our overview of quantum mechanics, the universe is fundamentally statistical in nature. This is reflected in the radioactive decay of an unstable isotope. It is impossible to tell exactly when one particular isotope will decay. However, a *group* of isotopes decay in a predictable and statistical rate, and they do so exponentially. We use the term, **half-life**, to describe this phenomenon, where the half-life is the time it takes for roughly one half of a group of isotopes to decay. Each isotope of an element has its own particular half-life for each type of decay, and it can range from well under one second to several billion years. For example, the half-life of Hydrogen-5 is about $8 \cdot 10^{-23}$ seconds, and the half-life of Uranium-238 is about 4.5 billion years, which is why we can easily find it on Earth. To find out how much of a given sample is left undecayed, we can use the following formula:

Fraction Remaining of Original Substance
$$= \left(\frac{1}{2}\right) \wedge \text{\# of Half-Lives that have passed}$$

So, after one half-life has passed, one-half of the original substance has remained, and the rest hast decayed into other isotopes and particles. After two half-lives, one-quarter of the original substance remains. Three half-lives mean that one-eighth remains, and so forth. You never end up with zero of the original isotopes remaining – just smaller and smaller fractions. This is shown in the figure below. Each half-life is marked as well.

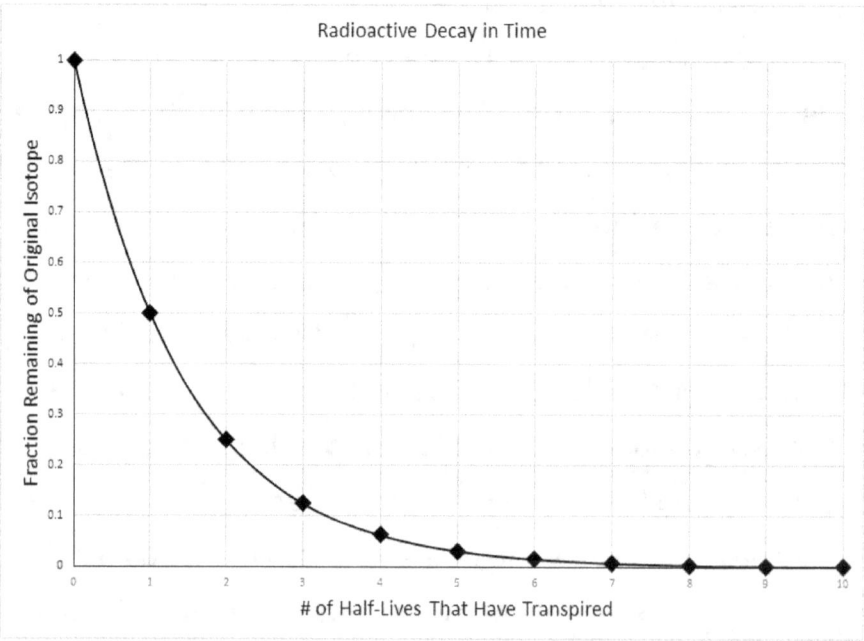

Sometimes, the byproducts of radioactive decay may also be radioactive. In those cases, there will be a chain of decay processes for the parent isotope until the daughter nucleus becomes a stable isotope.

Measuring Decay Strength

Different isotopes can undergo different decay processes and at different rates. Alpha particles, beta particles, gamma rays, x-rays, and other radioactive particles all interact with other particles differently as well. All these factors need to be considered for determining the strength of radiation. This also leads to an assortment of units to evaluate slightly different things.

First, we can measure the radioactivity -- the number of particles decaying or being transmitted. This is measured in becquerels, which is disintegrations per second. Curies are another unit for this, where 1 curie is $3.7 \cdot 10^{10}$ becquerels.

Second, we can look at the exposure, or the radioactivity through some environment. This is evaluated in terms of some equivalent charge (coulombs) divided by some amount of mass (kilograms). Historically, the units of roentgen are used, where one roentgen is $2.58 \cdot 10^{-4}$ C / kg.

The third way we look at radiation is the absorbed dose – the amount of radioactivity actually absorbed by that environment. This is measured in terms of energy (Joules) divided by that same mass (kilograms). The two named units are grays, Gy, where 1 Gy = 1 J/kg, and rads, where 1 rad = 0.01 J/kg = 0.01 Gy. rad stands for radiation absorbed dose.

Again, different kinds of radiation interact differently with matter. On an energy per decay basis, alpha radiation tends to be the least dangerous and gamma radiation tends to be the most hazardous. However, the intensity of radiation is critical – a little bit of gamma radiation is fine, but ingesting a powerful alpha emitter could be lethal. Thus, there is a need to gauge the medical effects of the absorbed dose of radiation. This is measured in sieverts, Sv, which has SI units of J/kg, but contains additional meaning as far as probabilities for cancer and tissue damage. Another unit is roentgen equivalent man, or rem, where 1 rem = 0.01 Sv.

The first medical weighting factor is the equivalent dose, which weights the absorbed dose based on the kind of radiation occurred. The absorbed dose gets multiplied by a radiation factor, W_R. If the radiation does not hit all parts of the body uniformly, a second weighting factor, the tissue factor, W_T, is included, and the sievert represents the effective dose. For most people, the effective dose from background sources is about 3 mSv (milli-sievert) per year, and about 100 mSv per year is the lowest level for a demonstrable increase in cancer risk. More information about the effective dose from different medical diagnostic procedures can be found at https://www.radiologyinfo.org/en/info/safety-xray and https://www.epa.gov/radiation/radiation-sources-and-doses. A useful comparison for effective dose sources can be found at https://xkcd.com/radiation/. This includes a very unofficial but educational way to think about health effects of radiation on us – the Banana Equivalent Dose (BED), the average dose of radiation from eating a banana!

Fusion and Fission

In addition to radioactive decay, there are two other interactions that can change the nucleus: fusion and fission. **Fusion** is where two or more nuclei or particles combine into a larger particle or nucleus. In the case of two nuclei fusing, the particles must be able to overcome the electromagnetic repulsion between them, since both nuclei are positively charged. The most common example of fusion is hydrogen fusing into helium in stars, which can happen due to the high temperatures and pressures present. Fusion can also be reproduced in a lab, but it requires extreme electromagnetic fields and a tremendous amount of energy to create the necessary environmental conditions, so fusion is not yet a practical source of energy for a power plant.

Fission is the other kind of nuclear reaction, where a heavy nucleus or particle is bombarded with other particles, and the original nucleus breaks into several pieces. The most common example is uranium nuclei being bombarded by free neutrons and releasing tremendous amounts of energy. This occurs naturally as well as in power plants, as certain uranium deposits show evidence that some amount of fission occurred in the past.

In a nuclear reactor in a power plant, the goal is to have a **stable chain reaction**, where one fission reaction produces exactly one other fission reaction. If it is less than a one-to-one ratio, the fission reactions will eventually stop and the reactor would need to be restarted. If more than one fission reaction occurs, more energy would be released than the system could potentially handle[*]. You never have to worry about a nuclear reactor exploding like a nuclear bomb – the physics isn't there, since nuclear weapons require very specific setups that cannot exist in a conventional reactor. Instead, pieces of the reactor overheat, causing mechanical failures and additional fission reactions, leading to radiation potentially spreading to other parts of the power plant.

Most nuclear reactors have four main components: fuel, coolant, moderator, and control rods. Sometimes, a material serves two purposed in the reactor. First, the reactor requires some fissile material, which is normally an isotope of Uranium, Plutonium, or Thorium. They each require somewhat different setups. The byproducts of the

---

[*] Slightly more than one reaction on average is necessary while the reactor powers up.

fission reactions can be extracted to serve additional industrial uses – it is not all waste. For some reactors, the fission of the fuel can produce another fissile isotope that can be used in a different reactor set-up – these are called breeder reactors.

The fission reactions produce a lot of heat due to the high kinetic energies of the products. That heat energy is transferred via a coolant to drive a turbine, thus creating our electrical power. The coolant is also needed to keep things from overheating. Typically, water is used as a coolant for both the fuel and the turbine, but a lead-bismuth mixture can be used in cooling the actual fissile material since it liquifies between 125 °C and 200 °C, depending on the mixture.

The last two components are there to regulate the reactions so that they remain as stable chain reactions. Moderators are used to slow the free neutrons that induce the fission reactions. For example, fission for Uranium-235 occurs primarily when the neutrons are slow moving. Often, moderators and coolants are the same substance. Water can be used a moderator, as can carbon in the form of graphite, liquid metals, and molten salts. The fast-moving neutrons produced by the fission reactions collide with the carbon atoms, transferring some of their kinetic energy. Control rods safely absorb free neutrons. Boron and indium are two examples of them. When a neutron collides with a control rod nucleus, the nucleus goes from one stable isotope to another.

Particle Physics and the Standard Model

We've discussed the main principles of quantum mechanics and its applications to light, electrons, and the nucleus. Again, the protons and neutrons in the nucleus are composite particles made up of quarks. The study of quarks and other subatomic particles is aptly called Particle Physics. Many popular science texts exist on various aspects of particle physics and cutting-edge research. Here, we'll just touch on a few key terms and ideas that have significant experimental backing.

The Standard Model is the term that we use for all that we more-or-less confidently know in particle physics and have been experimentally verified. String theory, for example, falls outside the Standard Model since we don't have the means to measure some of their predictions. In the Standard Model, we have a few basic categories of fundamental particles and four primary forces or interactions. Those four forces are gravity, electromagnetism, the weak nuclear force, and the strong nuclear force. Gravity is the least understood on the particle level, since it is too weak to be directly measured and cannot yet be fully meshed with quantum mechanics. We have already discussed electromagnetism in some detail. The strong force is what keeps the nucleus together, and the weak force is involved in beta decay. The strong and weak nuclear forces will be further explained as we discuss the different kinds of particles.

The first way we categorize fundamental particles is in how they exist in pairs or sets: they are either bosons or fermions. Two identical bosons can exist in the same quantum state. An example of a boson is light. Two light waves can have the same frequency, energy, polarization, and so forth. Two identical fermions, however, cannot share the same quantum state. An example of a fermion is an electron. Two electrons cannot have the same energy, angular momentum, or spin, which is why electron orbitals are formed. You can think of bosons as stackable cups and fermions as coffee mugs – two coffee mugs can't take up the same space.

Additionally, fermions *experience* forces or interactions, and bosons *transfer* the information about those forces or carry those forces from one fermion to another. Light is the boson that transfers the electromagnetic force. W^+, W^-, and Z bosons transfer information about the weak nuclear force, and gluons transfer information about the strong nuclear force. The Higgs Boson also factors into the weak nuclear force regarding the masses of the particles.

Fermions can be further categorized into quarks or leptons. Leptons do not interact via the strong nuclear force, and they include electrons and neutrinos. There are heavier versions of electrons called muons and tauons that have very short half-lives. Quarks do interact via the strong nuclear force. Additionally, quarks never exist in isolation. They can exist as quark and anti-quark pairs as mesons, three quarks join to be baryons, and very, very short-lived combinations of four or more quarks. The only stable quark combination is the proton. Free neutrons have a half-life of about 10 minutes, and are only stable in the nucleus.

The strong and weak nuclear forces are quite different from the other forces we have talked about. In electromagnetism, we only have positive and negative charges, so two varieties. Quarks have three possible "colors," red, blue, or green; and six possible "flavors," up, down, strange, charge, top, and bottom. Additionally, each quark has its own anti-quark, so there is a total of 36 of them. The weak nuclear force involves one of the quarks changing its flavor while interacting with leptons, and has some unique features regarding symmetry. Gluons, which carry information about the strong nuclear force, have a combination of "colors," and come in eight different varieties. Unlike the electromagnetic force, the strong nuclear force between two quarks increases with distance, so as the quarks get separated, it is more energetically favorable for a quark and anti-quark pair to pop into existence, which is why individual quarks cannot be isolated. As you can probably gather, things get very complicated very quickly, and most computations regarding the strong nuclear force require some heavy computer use. Unless working in a field that requires its knowledge, most physicists do not study the two nuclear forces in much detail – it requires a lot of time and effort to fully grasp the mathematics, and they don't really factor into anything above the subatomic level. For more information about basic particles, check out https://www.particleadventure.org.

This is all very abstract and esoteric stuff. However, when doing basic science research, you never know how it all may be applied in the future. This is definitely the case regarding particle physics and its eventual uses in medicine. Going back to X-rays in the early 20th century, research into particle physics has led to the development of powerful diagnostic tools and treatments. X-rays, CAT scans, MRIs, PETs, Proton Therapy, and Radiation Therapy are all byproducts of this basic science research. Medical Physics is its own expansive field, complete with its own sets of degree paths from the technician to the Ph.D. level!

Where Next?

You now have a rough overview of the sort of things Physicists learn and talk about. If you would like to get more into the math of things, I have also written "The Introductory Physics Workbook" to get you started. OpenStax College (https://openstax.org/) has free textbooks available that go along with introductory physics courses – one at the high school level ("Physics"), one set for a college algebra series ("College Physics"), and another set for a calculus-based series ("University Physics"). From there, there are specialized textbooks on each subject. Another worthwhile series is "The Theoretical Minimum" by Leonard Susskind. Lastly, on the philosophical side of the things, I strongly recommend "The Evolution of Physics" by Albert Einstein and Leopold Infeld, which was also written for a general audience.

www.ingramcontent.com/pod-product-compliance
Lightning Source LLC
Chambersburg PA
CBHW052314220526
45472CB00001B/106